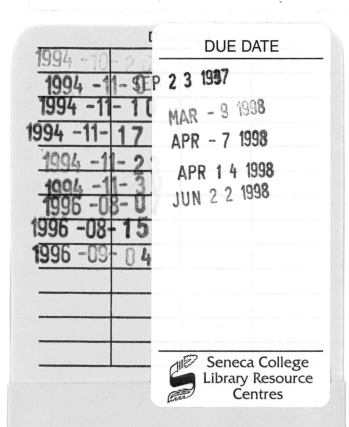

Clinical Opper

CLINICAL OPTICS

SECOND EDITION

Andrew R Elkington
MA, FRCS, FCOphth
Consultant Surgeon
Southampton Eye Hospital
Professor of Ophthalmology
University of Southampton

Helena J Frank
BMedSci, FRCS, FCOphth
Consultant Ophthalmologist
Royal Victoria Hospital, Bournemouth

OXFORD

BLACKWELL SCIENTIFIC PUBLICATIONS

LONDON EDINBURGH BOSTON

MELBOURNE PARIS BERLIN VIENNA

© 1984, 1991 by
Blackwell Scientific Publications
Editorial offices:
Osney Mead, Oxford OX2 0EL
25 John Street, London WC1N 2BL
23 Ainslie Place, Edinburgh EH3 6AJ
3 Cambridge Center, Cambridge,
  Massachusetts 02142, USA
54 University Street, Carlton,
  Victoria 3053, Australia

Other Editorial Offices:
Arnette SA
2, rue Casimir-Delavigne
75006 Paris
France

Blackwell Wissenschaft
Meinekestrasse 4
D-1000 Berlin 15
Germany

Blackwell MZV
Feldgasse 13
A-1238 Wien
Austria

First Edition published 1984
Reprinted 1988
Second Edition published 1991

Set by DP Photosetting, Aylesbury, Bucks
Printed and bound in Great Britain by
Hartnolls Ltd, Bodmin, Cornwall

DISTRIBUTORS
  Marston Book Services Ltd
  PO Box 87
  Oxford OX2 0DT
  (*Orders:* Tel: 0865 791155
          Fax: 0865 791927
          Telex: 837515)

USA
  Mosby-Year Book, Inc.
  11830 Westline Industrial Drive
  St Louis, Missouri 63146
  (*Orders:* Tel: 800 633-6699)

Canada
  Mosby-Year Book, Inc.
  5240 Finch Avenue East
  Scarborough, Ontario
  (*Orders:* Tel: 416 298-1588)

Australia
  Blackwell Scientific Publications
  (Australia) Pty Ltd
  54 University Street
  Carlton, Victoria 3053
  (*Orders:* Tel: 03 347-0300)

British Library
Cataloguing in Publication Data
Elkington, Andrew R.
  Clinical Optics.—2nd. ed.
  1. Optics
  I. Title   II. Frank, Helena J.
  535.0246177

ISBN 0–632–03139–5

Contents

Preface

We have written this book in the hope of helping those trainee Eye Surgeons who are preparing to take their basic profession examinations. We have assumed that the optics they learned at school has long since been forgotten and we have therefore set out to explain the subject from first principles. Our aim has been to keep the text logical and simple and we have used diagrams liberally to complement the written word. These diagrams have themselves been simplified so that they can be both easily memorised and reproduced—even under the stress of examination conditions.

Throughout the book we have emphasized the clinical relevance of each topic and we trust that this approach will allow the reader to understand better the optical problems experienced by many of his patients.

ACKNOWLEDGEMENTS

Many people have helped and encouraged us during the evolution of this book and we are grateful to all of them. We thank particularly those trainee Eye Surgeons working in Wessex who first suggested to us the need for such a text. We are grateful to both Mr. Edward Zorab and Mr. Christopher Dean Hart who read early drafts of some of the chapters and who urged us to continue writing. The first edition owed much to Claire Little who prepared our drawings for press and Sue Emery who produced the typescript; Mina Yuille kindly made up the index and Patrick Trevor Roper wrote a generous foreword. In preparing this Second Edition we have been greatly helped by Mr. Jock Anderson. Professor John Marshall and Dr. Alan McKenzie kindly helped us with the section on lasers. Finally the book would never have reached its readership without the efforts of Bridget Cook, Richard Zorab and Richard Miles of Blackwells and it is a pleasure to thank them for all their support.

Chapter 1   Properties of light

Light may be defined as energy to which the human eye is sensitive. Scientists do not yet fully understand the true nature of light in the physical sense, but the behaviour and properties of light have been extensively studied and are well known.

This book aims to describe with the aid of diagrams those aspects of optics which are relevant to the practising ophthalmologist. In this first chapter a simple account is given of the nature and properties of light.

Electromagnetic spectrum: optical radiation: colour

Optical radiation lies between X-rays and microwaves in the electromagnetic spectrum (Fig. 1.1), and is subdivided into seven wavebands. Each of these seven wavebands group together wavelengths which elicit similar biological reactions. These seven domains are ultraviolet C (UV-C), 200–280 nm; ultraviolet B (UV-B), 280–315 nm; ultraviolet A (UV-A), 315–400 nm; visible radiation, 400–780 nm; infrared A (IRA), 780–1400 nm; infrared B (IRB), 1400–3000 nm, infrared C (IRC), 3000–10000 nm. As with all electromagnetic radiation, the shorter the wavelength, the greater the energy of the individual

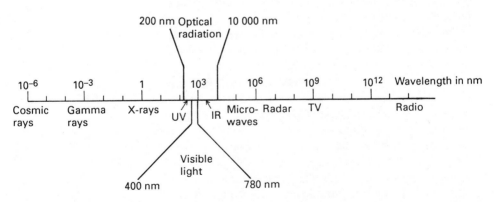

Fig. 1.1 The electromagnetic spectrum.

The normal eye is able to discriminate between light of shorter or longer wavelength within the visible spectrum

1

quanta, or *photons*, of optical radiation.

The cornea and sclera of the eye absorb essentially all the incident optical radiation at very short wavelengths in the ultraviolet (UV-B and UV-C) and long wavelengths in the infrared (IR-B and IR-C). The incident UV-A is strongly absorbed by the crystalline lens while wavelengths in the range 400–1400 nm (visible light and near infrared), pass through the ocular media to fall on the retina. The visible wavelengths stimulate the retinal photoreceptors giving the sensation of light while the near infrared may give rise to thermal effects. Because the refractive surfaces of the eye focus the incident infrared radiation on the retina, it can cause retinal damage, e.g. eclipse burns.

The normal eye is able to discriminate between light of shorter or longer wavelength within the visible spectrum by means of colour sense originating from three different classes of cone cells. When viewing a band of light composed of the whole visible spectrum arranged in order of decreasing wavelength the familiar rainbow colour pattern is seen. Light of the longest wavelength is seen as red, progressing through orange, yellow, green, blue and indigo to the violet of the shortest visible wavelengths. White light is a mixture of the wavelengths of the visible spectrum.

The retinal photoreceptors are also sensitive to wavelengths between 400 nm and 350 nm in the near ultraviolet (UV-A). These wavelengths are normally absorbed by the lens of the eye. In aphakic eyes or pseudophakic eyes with intraocular implants without UV filter such UV radiation gives rise to the sensation of blue or violet colours. Newly aphakic patients frequently remark that 'everything looks bluer than before the operation'.

Of greater concern is the recent evidence that wavebands between 350 nm in the UV and 441 nm in the visible are potentially the most dangerous for causing retinal damage under normal environmental conditions. Intraocular implant lenses are made of polymethylmethacrylate, which only absorbs UV light below 320 nm. The retina of the aphakic or pseudophakic eye is therefore deprived of the natural protection afforded by the crystalline lens. Intraocular implant lenses are therefore being produced which incorporate a UV-A absorbing substance and in future may also be made to contain a yellow pigment, like the natural lens, and thus also block light between 400 nm and 441 nm.

Wave theory of light

The path of light through an optical medium, e.g. glass is always straight if no obstacle or interface between optical media is encountered. Diagrammatically light is represented as a straight arrowed line or ray. (By tradition optical diagrams show rays travelling from left to right on the page.) However, some experimental observations of the behaviour of light are not fully explained by the simple concept of light as rays and it is now understood that light really travels as waves, although its path is often represented as a 'ray'.

Fig. 1.2 Light leaving a point source (a) Light represented as rays (b) Light represented as waves (c) Light represented as wave fronts.

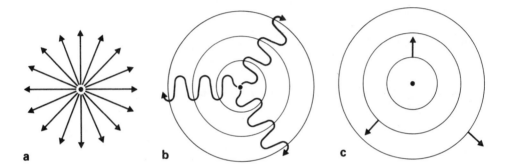

a   b   c

Fig. 1.2 illustrates the different ways of depicting the progress of light away from a point source. Fig. 1.2a shows the light as rays. Fig. 1.2b shows the wave motion of each ray while Fig. 1.2c illustrates the wave front set up by the combined effect of many rays, the concentric circles being drawn through the crests of the waves. The same effect is seen if a stone is dropped into still water. Viewed from above circular waves travel outwards from the point of impact (wave fronts in Fig. 1.2c). If the process were viewed in cross section the waves would appear as ripples travelling away from the centre of disturbance (wave motion in Fig. 1.2b).

Wave motion consists of a disturbance, or energy, passing through a medium. The medium itself does not move, but its constituent particles vibrate at right angles to the direction of travel of the wave (Fig. 1.3). (Imagine a ribbon tied to a rope along which a wave is 'thrown'. The crest of the wave moves along the length of the rope, but the ribbon moves up and down at one point on the rope.)

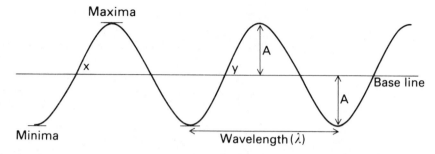

Fig. 1.3 Wave motion.

The *wavelength*, λ, is defined as the distance between two symmetrical parts of the wave motion. One complete oscillation is called a *cycle*, e.g. x y, Fig. 1.3 and occupies one wavelength. The *amplitude*, A, is the maximum displacement of an imaginary particle on the wave from the base line. Any portion of a cycle is called a *phase*. If two waves of equal wavelength (but not necessarily of equal amplitude) are travelling in the same direction but are 'out of step' with each other, the fraction of a cycle or wavelength by which one leads the other is known as the *phase difference* (Fig. 1.4).

Fig. 1.4 Wave motion: phase difference.

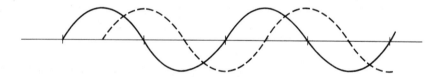

Fig. 1.4 shows two waves of equal wavelength which are out of phase by one quarter of a wavelength (phase difference equals 90°, the complete cycle being 360°).

Light waves that are out of phase are called *incoherent*, while light composed of waves exactly in phase is termed *coherent*.

Interference

When two waves of light travel along the same path, the effect produced depends upon whether or not the waves are in phase with one another. If they are in phase the resultant wave will be a summation of the two, and this is called *constructive interference* (Fig. 1.5a). If the two waves of equal amplitude are out of phase by half a cycle (Fig. 1.5b) they will cancel each other out, *destructive inter-*

ference. The final effect in each case is as if the waves were superimposed and added (in the algebraic sense) to each other. Phase differences of less than half a cycle thus result in a wave of intermediate amplitude and phase (Fig. 1.5c).

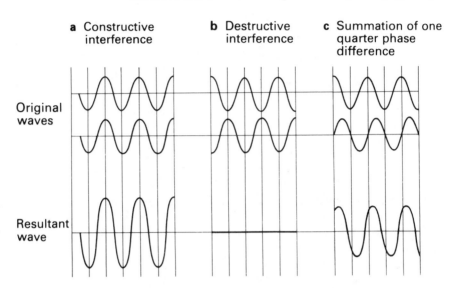

Fig. 1.5 Interference of two waves.

Destructive interference occurs within the stroma of the cornea. The collagen bundles of the stroma are so spaced that any light deviated by them is eliminated by destructive interference.

Interference phenomena are also utilized in optical instruments. One example is low reflection coatings which are applied to lens surfaces. The coating consists of a thin layer of transparent material of appropriate thickness. Light reflected from the superficial surface of the layer and light reflected from the deep layer eliminate each other by destructive interference.

Diffraction

When a wave front encounters a narrow opening or the edge of an obstruction (Fig. 1.6) the wave motion spreads out on the far side of the obstruction. It is as if the edge of the obstruction acts as a new centre from which secondary wave fronts are produced which are out of phase with the primary waves. This phenomenon is called *diffraction*.

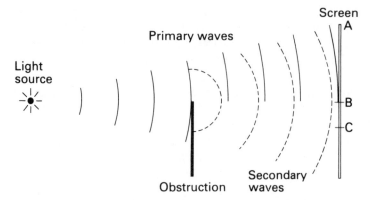

Fig. 1.6 Diffraction (exaggerated).

The intensity of the light falling on zone AB is reduced to some extent by interference between the primary and secondary waves. The light falling on zone BC is derived from secondary waves alone and is of much lower intensity.

When light passes through a circular aperture, a circular diffraction pattern is produced. This consists of a bright central disc surrounded by alternate dark and light rings. The central bright zone is known as the *Airy disc*.

Diffraction effects are most marked with small apertures, and occur in all optical systems including lenses, optical instruments and the eye. In the case of lenses and instruments, the diffraction effect at the apertures used is negligible compared with the other errors or aberrations of the system (see Chapter 8). In the case of the eye, diffraction is the main source of image imperfection when the pupil is small. However, the advantage of a large pupil in reducing diffraction is outweighed by the increased effect of the aberrations of the refractive elements of the eye (Chapter 8).

Limit of resolution; resolving power

The terms *limit of resolution* or *resolving power* refer to the smallest angle of separation (w) between two points which allows the formation of two discernable images by an optical system (Fig. 1.7). The limit of resolution is reached when two Airy discs are separated so that the centre of one falls on the first dark ring of the other.

Fig. 1.7 Angle of resolution (w).

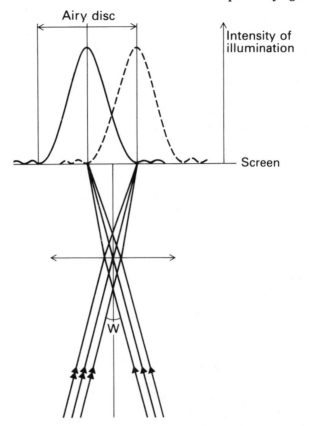

Tests of visual acuity—resolving power of the eye

It is important in clinical practice to be able to measure the resolving power of the eye. Many tests have been designed for this purpose. The underlying principle common to all tests is that the subject is required to recognize a shape which subtends a known angle at the eye when viewed from the appropriate distance. The normal limit of resolution for the eye is one minute of a degree. At this angle of separation the two retinal images are separated by at least one non-stimulated cone.

The visual acuity is commonly tested using the Snellen Test Type. A chart bears letters constructed so that each letter subtends a visual angle of 5 minutes (min) of a degree when viewed from the specified distance. Therefore, in order to recognize a letter the eye must have a limit of resolution of one minute of a degree.

The chart bears letters of diminishing size, the largest

having a viewing distance of 60 metres (m), with smaller letters for distances of 36 m, 24, 18, 12, 9, 6 and 5 m. The patient is usually positioned 6 m from the chart. A normal eye reads the 6 m letters from a distance of 6 m, and is said to have 6/6 vision. A weaker eye may only be able to resolve the larger letters, e.g. the 36 m size, and is said to have 6/36 vision. If, for any reason the patient reads the chart from a different distance, the numerator of the acuity is amended accordingly, e.g. if the test is done at 5 m the above examples become 5/6 and 5/36 vision.

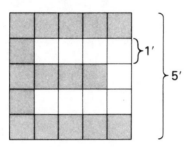

Fig. 1.8 Snellen Test Type letter.

Polarization of light

The orientation of the plane of the wave motion of rays comprising a beam of light is random unless the light is polarized. Fig. 1.9a shows a beam cut across and viewed end-on—the light is travelling perpendicular to the page.

a Non-polarized light

b Polarized light

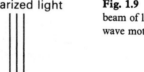

Fig. 1.9 Cross section of beam of light to show plane of wave motion.

In contrast, Fig. 1.9b shows the cross section of a beam of light in which the individual wave motions are lying parallel to each other. Such a beam is said to be *polarized*.

Polarized light is produced from ordinary light by an encounter with a polarizing substance or agent. Polarizing substances, e.g. polaroid plastic, only transmit light rays which are vibrating in one particular plane. Thus only a proportion of incident light is transmitted onward and the emerging light is polarized.

In nature, light is polarized on reflection from a plane surface, such as water, if the angle of incidence is equal to the polarizing angle for the substance. The polarizing angle is dependent on the refractive index of the substance comprising the reflecting surface (cf Chapters 2 and 3). At other angles of incidence the reflected light is partially polarized, i.e. a mixture of polarized and non-polarized light. Furthermore, the plane of polarization of the reflected light from such a surface is parallel with the surface. As most reflecting surfaces encountered in daily life are horizontal it is possible to prepare polarized sunglasses to exclude selectively the reflected horizontal polarized light. Such glasses are of great use in reducing glare from the sea or wet roads.

Applications of polarized light in ophthalmology include its use in instruments such as the slit-lamp and ophthalmoscope, to reduce reflected glare from the cornea; in the assessment of binocular vision polarizing glasses may be used to dissociate the eyes, e.g. in the Titmus test (p. 10); in pleoptics to produce Haidinger's brushes, and in optical lens making to examine lenses for stress.

Stereoscopic vision
Stereopsis is the ability to fuse slightly dissimilar images, which stimulate disparate retinal elements within Panum's fusional areas in the two eyes, with the perception of depth. It is graded according to the least horizontal disparity of retinal image that evokes depth perception, and is measured in seconds of arc.

The normal stereoacuity is approximately 60 seconds of arc or better (slightly different values are quoted by different workers). An individual with very good stereoscopic vision may have a stereoacuity of better than 15 seconds of arc, which is the smallest disparity offered in the Frisby stereotest (range 600–15 seconds of arc). The maximum stereoacuity is achieved when the images fall on the macular area of the retina, where the resolving power of the eye is at its best. Good stereoacuity is therefore a product of central single binocular vision. A stereoacuity of better than 250 seconds of arc is said to exclude significant amblyopia, while a stereoacuity of worse than 250 seconds of arc may be an indicator of amblyopia.

Clinical tests of stereoacuity

There are quite a variety of tests of stereoacuity available, but those most commonly used in the UK are the Titmus test, the TNO test, the Frisby test and the Lang stereotest.

The *Titmus test*, which includes the Wirt fly test, is in the form of vectographs. A vectograph consists of two superimposed views presented in such a way that the light entering the eye is plane polarised, the light from one view being at right angles to that from the other. The composite picture must be viewed through a polarizing visor or spectacles.

The Wirt fly is the largest target in the test, which also includes graded sets of animals and circles, one of which is disparate and appears to stand forward. The test must be viewed at 40 cm, and covers a range of stereoacuity from approximately 3000 to 40 seconds of arc.

The *Frisby test* consists of three clear plastic plates of different thicknesses. On each plate there are four squares filled with small random shapes. One square on each plate contains a 'hidden' circle, which is printed on the back surface of the plate. The random shapes give no visual clue to the edge of the 'hidden' circle, and the test is purely three-dimensional and does not require polarizing or coloured glasses to be worn. At a 40 cm viewing distance the plates show a disparity of 340, 170 and 55 seconds of arc, and by adjusting the viewing distance the test can be used to give a disparity range from 600 to 15 seconds of arc.

The *TNO test* comprises computer-generated random dot anaglyphs. An anaglyph is a stereogram in which two disparate views are printed in red and green respectively on a white ground. Red-green spectacles are worn to view the anaglyph. The eye looking through the red filter sees only the green picture, as black, and the eye looking through the green filter sees the red picture, again as black, and the two views may be fused to give a stereoscopic effect. In the TNO test the disparities range from 480 to 15 seconds of arc.

The *Lang stereotest* targets are made up of fine vertical lines which are seen alternately by each eye when focused through built-in cylindrical lens elements. The displacement of the random dot images creates disparity ranging from 1200 to 550 seconds of arc. The test card must be held parallel to the plane of the patient's face to avoid giving uniocular clues. The test is viewed at a normal reading distance.

Photometry

Photometry is the quantitative measurement of light. That is to say, how much light is being emitted (luminous flux, luminous intensity), or falling on a surface (illumination) or being reflected from that surface (luminance). The eye is the original photometer, and it is most sensitive to yellow–green light, its sensitivity declining towards both ends of the visible spectrum. Thus, when viewing a spectrum, the yellow–green portion looks 'brighter' than the red–orange or blue–violet bands. Electrical photometers are designed so that their sensitivity to different wavelengths mimics that of the eye. The definitions and units used in photometry are illustrated in the following diagrams (Figs. 1.10–1.12). The units used can be confusing as several systems of nomenclature are in use. The commonly used units are described below.

The total flow of light in all directions from a source is termed the *Luminous flux* of the source. Luminous flux is measured in *lumens*.

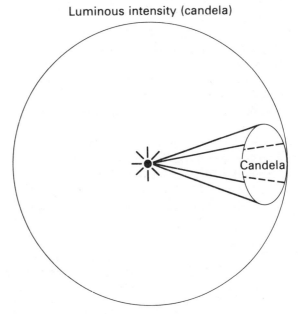

Luminous intensity (candela)

Luminous flux (lumen)

Fig. 1.10 Units of light emitted by a source of light: luminous flux and luminous intensity.

The light emitted in a given direction is termed the *Luminous intensity* of the source in the given direction. The unit of luminous intensity is the *candela*. The size of the cone of light is measured in steridians, the steridian being the unit of solid angle.

1 candela = 1 lumen per steridian.

Luminous intensity is sometimes referred to as *candle-power*.

The candela is the modern unit of luminous intensity and its definition is based on a standard electrical filament lamp. The old unit was the candle, based on a standard wax candle. The two units are equal to one another and the terms candela and candle are interchangeable.

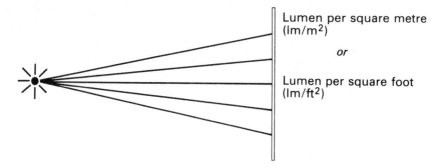

Lumen per square metre
(lm/m^2)

or

Lumen per square foot
(lm/ft^2)

Illumination refers to the light arriving at a surface. Surface *illumination* is measured in *lumen per square metre* (lumen/m^2) or *lumen per square foot* (lumen/ft^2). (The old names for these units were the metre-candle also called lux, and the foot-candle respectively.)

The illumination of a surface decreases the further it is from the light source. The surface illumination is inversely related to the square of the distance of the surface from the source (the inverse square law).

The illumination of a surface is also dependent upon the angle of the incident light to the surface. The illumination is directly related to the angle of incidence (the cosine law).

Fig. 1.11 Surface illumination.

Thus $E = \dfrac{I.\cos i}{d^2}$

where E is the illumination
I is the luminous intensity
i is the angle of incidence
d is the distance between source and surface.

Luminance

Luminance is the measure of the amount of light reflected or emitted by a surface. There are two ways of measuring luminance and therefore two sets of units in use (Fig. 1.12).

In the so-called 'Lambert system' (Lambert 1727–77) the surface is assumed to be a perfect diffuser—that is, to appear equally bright from any direction of view. Few perfectly diffusing surfaces exist but 'matt' surfaces fulfil the requirement when the angle of incidence is small and the direction of viewing is close to the normal (see p. 22). Assuming the surface to be a perfect diffuser, luminance is measured as total luminous flux emitted by unit surface area. *A surface emitting or reflecting 1 lumen/cm²* *has a luminance of 1 lambert.* However, the unit most commonly used is the foot-lambert which is the luminance of a surface reflecting or emitting 1 lumen/ft².

Fig. 1.12 Units of luminance.

a The foot lambert

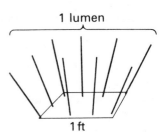

1 lumen

1 ft

b One candela per square foot

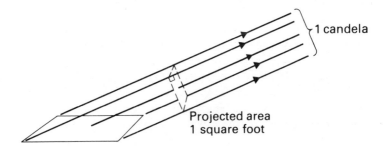

1 candela

Projected area
1 square foot

The other method of measuring luminance specifies the direction of view. The surface area considered is not the actual emitting or reflecting surface but the area of that surface when projected on a plane perpendicular to the direction of view (Fig. 1.12b). Because the direction of view is specified, the candela replaces the lumen as the unit of light. The units are thus the candela/cm² or the candela/ft², etc. for the various measures of unit area.

The two sets of units are related to each other as follows:

1 candela/ft² = 1 foot-lambert.

Lasers

It seems appropriate to conclude this chapter with a simple account of how lasers work. The word LASER is an acronym from their mechanism of action—**L**ight **A**mplification by **S**timulated **E**mission of **R**adiation.

At the atomic level, quanta or *photons* of light energy are emitted from atoms whose electrons fall from a higher to a lower energy level. This process occurs in a random fashion in ordinary light sources so that the light is emitted in all directions and is composed of a mixture of wavelengths whose wavefronts are out of phase (incoherent).

A laser consists of an energy source, usually light, which is 'pumped' into a transparent laser material to excite the atoms to high energy levels. One of these, the upper laser energy level (Fig. 1.13) must have a comparatively long lifetime to allow a population of excited atoms to accumulate. When there are more atoms in the excited state than in the lower energy level or ground state, *population inversion* is said to have occurred.

An atom in the high energy or excited state may be stimulated to emit light by an encounter with a photon if the photon is of the same wavelength as that which the atom would naturally emit. The photon produced by this *stimulated emission* is in the same direction and in phase with the stimulating photon, and the atom falls to the lower laser energy level (Fig. 1.13). The lower laser level must have a short lifetime allowing the atoms to decay back to the ground state and enter the cycle afresh. However, because heat is produced during laser operation, the lower laser level must be far enough above the ground state to be inaccessible to merely heated atoms. If the lower laser level became heavily populated by thermal pumping, the lasing process would be blocked.

Fig. 1.13 Energy levels of a simple laser.

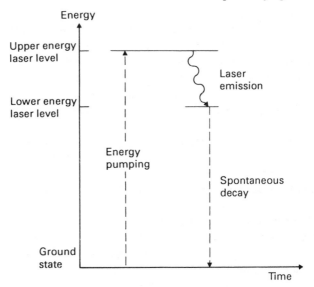

In a gas laser, the lasing material is housed in a tube which has a mirror at each end (Fig. 1.14). The distance between the mirrors must equal a multiple of the wavelengths of the light emitted so that resonance can occur. When a photon encounters an excited electron and stimulated emission occurs, the light emitted travels down the tube, is reflected and rereflected at both mirrors. Because the mirrors are precisely aligned and a whole number of wavelengths apart, the light which has traversed the tube is still exactly in phase with itself on its second and subsequent journeys. Thus it reinforces itself. This is known as *resonance*. Meanwhile other stimulated emissions are taking place so that the light traversing the tube gets stronger and stronger while remaining exactly in phase (coherent) and the lasing process is under way.

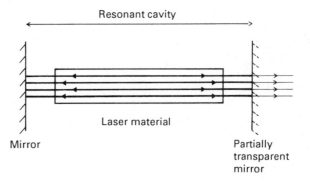

Fig. 1.14 Laser tube.

If one of the mirrors is made partially transparent, some of the light may be allowed to leave the tube. This light will be coherent (the wavefronts in phase), monochromatic (of one wavelength) and collimated (all the rays parallel). Light is produced continuously and such a laser is said to be operating in *continuous-wave (CW) mode*.

The actual luminous flux emitted by a laser is relatively small (lasers are very inefficient in that a great deal of energy has to be 'pumped' into them in order to maintain the lasing process). However, because the luminous flux is not scattered in all directions but is concentrated in a fine parallel beam, the beam of light is exceedingly bright. A laser producing approximately 5 lumens of light may have a beam of luminous intensity 500 million candela. Another useful comparison is that a 1 Watt laser produces a retinal irradiance 100 million times greater than that of a 100 Watt incandescent bulb.

Q-switched mode of laser operation

The time characteristics of laser emissions vary depending on the laser material, the mode of excitation and the design of the laser.

If a time-variable absorber or Q-switch is incorporated into the laser tube it is possible to change the laser output from CW mode to single, very brief, very high power pulses. This is called the *Q-switched mode* of operation. (Q stands for Quality factor which is an engineering term related to the ratio of energy storage to energy loss in the laser medium. The higher the Q, the more energy is stored.)

Exciting energy is pumped into the laser with the absorber in place (Fig. 1.15). The ions of the laser material

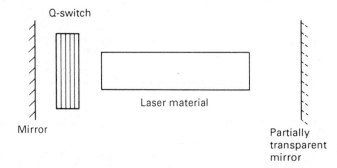

Fig. 1.15 Q-switched or mode-locked laser.

are raised to high energy levels but, because the absorber blocks one of the mirrors, resonance and loss of the upper laser level by stimulated emission of light do not occur. Very high levels of population inversion can thus be obtained. If the absorber is suddenly made transparent, resonance occurs and a single very high power pulse of laser light is emitted as the large population of excited ions are stimulated to discharge their energy. Q-switched pulses are of the order of 10–50 nanoseconds (10^{-9} of a second) duration. Various absorbers are used including rotating mirrors, saturable dye cells and electrooptic switches.

Mode locked laser operation

Even laser light is not pure, in the sense of being a single wavelength, for the following reasons. The length of the laser tube, often of the order of 1 metre, is enormous compared to the wavelength of laser light. It is therefore possible for multiples of several wavelengths to 'fit' into the tube length. In the case of solid-state lasers, the heat generated during operation may cause expansion of the laser crystal, altering the distance between the mirrors. Gas lasers, on the other hand, have wavelength impurities due to the Doppler effect. The ions of the gas are moving randomly within the tube, and the exact wavelength of each stimulated emission depends on whether the emitting ion is travelling in the same or opposite direction as the emitted photon.

In practice, the *line width* of emitted laser light can vary from less than 0.001 nm to 10's of nanometres around the main wavelength, depending on the laser material in use. The various 'sub-wavelengths' or modes operate independently and are not in phase with one another. This state of affairs is called the *free-running mode*.

A further refinement of Q-switching is the production of a train of laser pulses separated from each other by a specific time interval and with all the 'sub-wavelengths' or modes in phase, i.e. *mode locked*.

The laser construction is the same as for Q-switched operation (Fig. 1.15) but the distance between the mirrors is longer. The absorber is only inactivated very briefly at a pulse repetition frequency (PRF) equal to the time needed for a photon of light to make the round trip of the laser tube. Thus, short, high power pulses of light are emitted separated by intervals equal to the time taken by the light to make the round trip. A bleachable dye is used as the absorber in ophthalmic lasers. The dye remains opaque until the laser light grows strong enough to saturate it when it becomes transparent for a brief instant, allowing release of the pent-up laser energy. The pulse of laser light bleaches the dye every round trip of the laser tube. However, light arriving at the Q-switch out of phase with the main pulse is absorbed, and so all wavelengths (modes) are compelled to travel in phase, i.e. mode locked. The combined energy of all the modes summates to produce a brief, exceedingly powerful pulse. The broader the line width of a laser (i.e. the more 'sub-wavelengths' present) the briefer and more powerful the mode locked pulse produced. The pulses are released in *pulse trains* of up to 12 pulses after which there is a longer interval before the next pulse train begins.

For the neodymium-YAG laser which has a comparatively broad line width, the mode locked pulses are of 30 picoseconds (10^{-12} second) duration, separated by 5–7 nanoseconds (10^{-9} second) with 7–10 pulses per train. In the Q-switched mode single pulses lasting 10–20 nanoseconds are produced. However, the total energy released in the two modes of operation is approximately equal.

Lasers used in ophthalmology

Laser technology is advancing very fast, with several new lasers being developed each year which may have clinical potential. The use of lasers is therefore one of the most rapidly advancing frontiers of ophthalmology. The lasers described below have an accepted place in ophthalmic practice.

The *argon laser* is in everyday use in ophthalmology. It is one of the class of noble gas ion lasers, the other in ophthalmological use being the krypton laser. The argon laser produces a blue-green light composed of a mixture of 70% 488 nm (blue) and 30% 514.5 nm (green) light.

The *krypton laser* which produces two strong lines in the visible spectrum, 568 nm (yellow) and 647 nm (red), is also in clinical use.

The *helium-neon(He-Ne) laser* is a low-power gas laser producing a red beam used as an aiming beam for lasers such as the neodymium-YAG and diode lasers whose output is outside the visible spectrum.

Dye lasers use one of many liquid materials available as the laser material. The output of a dye laser can be varied over a range of wavelengths, and by changing the mirror configurations a different range of wavelengths can be obtained from the same machine.

Diode lasers have sprung from the recent advances in semi-conductor technology. The laser crystals are as small as a single grain of salt, and emit in the 790–950 nm range (infrared) which is invisible to the human eye. A visible aiming beam must therefore be incorporated in the instrument. The diode laser operates in continuous wave (CW) mode, but is a power on demand system, that is, the lasing process only occurs when power is switched on. Diode lasers are extremely compact, highly efficient and relatively inexpensive.

The *neodymium-yttrium-aluminium-garnet (Nd-YAG) laser* is a powerful continuous wave (CW) laser which is usually used Q-switched when treating the eye. Neodymium ions produce the laser light and they are contained as impurity ions within an optically 'pure' YAG crystal. In this way a much higher concentration of active laser (Nd) ions can be achieved than in a gas laser medium. The laser output is in the infrared at 1.065 µm and invisible to the eye. A red He-Ne laser aiming beam is therefore incorporated into the instrument.

Light from the CW lasers mentioned above can be delivered by a fibreoptic system built into a slit-lamp microscope, or in the case of the Nd-YAG laser, via a system of prisms. A contact lens containing mirrors or incorporating a lens is used to focus the laser light onto the target structure, (cf page 35, gonioscopy and 3-mirror lenses; page 167, fundus viewing contact lens). Both the direct and the indirect ophthalmoscopes have been used as delivery systems and laser light may also be delivered to the retina at surgery via an intraocular fibreoptic probe.

Another recent addition to the range of lasers finding application in ophthalmology are the *excimer lasers*. 'Excimer' is as term derived from 'excited dimer', a dimer being a molecule composed of two atoms of the same species (homonuclear) or different species (heteronuclear). In the excited state the excimer is bound, but on decaying to the ground state the molecule rapidly dissociates into its constituent atoms, facilitating population inversion.

The *argon-fluoride excimer laser* is currently finding use as a corneal cutting device. This laser's output is at 193 nm (ultraviolet) and at this short wavelength each photon is of very high energy, more than 6 electron volts. (The energy of a carbon-carbon bond is approximately 3 electron volts while the energy of a Nd-YAG infrared photon is just over 1 electron volt.) The high energy excimer photons are able to break the intramolecular bonds of the corneal surface tissue enabling a fine layer to be removed with each pulse without thermal damage to the remaining cornea. This process is known as *photoablation*. By manipulating the characteristics of the laser beam the cornea may be sculptured to a desired curvature.

Mode of action of ophthalmic lasers

Laser energy can produce various effects in the eye, depending on the wavelength used and the pulse duration, i.e. the mode of delivery. The mechanism of effect may be thermal, ionizing or by photoablation. Lasers are also used to incise other tissues by a vaporising effect.

Thermal effect. Light energy of different wavelengths is selectively absorbed by the various pigments within the eye. For example, the xanthophyll pigment found at the macula absorbs the blue (488 nm) argon laser light strongly and the green (514.5 nm) to a lesser extent. In contrast to this, the wavelengths emitted by the krypton laser (568 nm and 647 nm) are very little absorbed. This is the reason why the krypton laser is preferred in the treatment of macular disorders. When energy is delivered relatively slowly, i.e. in CW mode, a heating effect is achieved and localised thermal burns may be inflicted on the target structure. Accurate thermal burns may also be applied to the iris and trabecular meshwork. The argon, krypton, dye and diode lasers may be used in CW mode to achieve a thermal effect.

Ionizing effect. If a large amount of energy is rapidly delivered into a tiny focal volume in a very short time, the constituent atoms of the target material will be ionized and disintegrate into a collection of ions and electrons called a plasma. This rapidly expands causing acoustic and shock waves which can disrupt tissue in the immediate neighbourhood of the blast. The Q-switched or mode-locked Nd-YAG laser is used in this way to disrupt structures within the anterior part of the eye such as the lens capsule, posterior synechiae or to make a peripheral iridotomy. It has also been used in the posterior segment to cut vitreous strands, but damage may be done to the retina or crystalline lens if the focus is too close to them.

Photoablation. This is the mode of action of the argon-fluoride excimer laser and is described above.

Vaporising effect. If the tissue temperature is raised to 100°C, vaporisation of the tissue fluid occurs. By this mechanism an almost bloodless incision can be made, and this is of great use in other branches of surgery where the carbon dioxide laser is used for this purpose. Its output is at 10.6 μm, and 90 per cent of it is absorbed within a thickness of 200 μm of tissue. In other words, the energy is primarily absorbed by the surface cells which are vaporised but thermal conduction damages adjacent tissues. Because the energy is absorbed at the tissue surface, it cannot be transmitted via the ocular media and must be directly applied to its target. It is yet to find a use in ophthalmic practice.

Photochemical effect. In this case the damage results from photochemical reactions in which single photons of specific wavelength convert individual absorbing molecules into one or more different molecules (free radicles) which may be toxic to the cell. The use of low energy laser radiation in conjunction with haematoporphyrin derivative (HPD) in the treatment of choroidal malignant melanoma is being explored.

Laser safety

All surgical lasers, including those used in ophthalmology, are by definition capable of causing damage to the eye, and are classified as Class 4 laser systems. It is essential that their effects be limited to the therapeutic end desired, and that accidental damage is not inflicted on the patient, the surgeon or other persons in the vicinity. To this end there are strict safety regulations which must be adhered to.

Where there is a risk to the surgeon from reflected laser light, protection is provided by shutters or filters built into the instrument which operate when the laser is fired. The filters absorb the laser light but transmit enough light of other wavelengths to allow the effect of the laser on the target to be observed. The laser area should be free of personnel but if attendance of other individuals is essential protective goggles should be used.

Reflection of Light   Chapter 2

When light meets an interface between two media, its behaviour depends on the nature of the two media involved. Light may be absorbed by the new medium, or transmitted onward through it (see Chapter 3), or it may bounce back into the first medium. This 'bouncing' of light at an interface is called *reflection*. It occurs, to some degree, at all interfaces even when most of the light is transmitted or absorbed. It is by the small amount of reflected light that we see a glass door and thus avoid walking into it.

Laws of reflection

The following laws govern reflection of light at any interface and are illustrated in Fig. 2.1.
1. The incident ray, the reflected ray and the normal to the reflecting surface all lie in the same plane.
   (The 'normal' is a line perpendicular to the surface at the point of reflection).
2. Angle of incidence, i, equals the angle of reflection r.

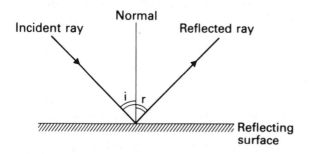

Fig. 2.1 Reflection at a plane surface.

Reflection at an irregular surface

When parallel light encounters an irregular surface, it is scattered in many directions (Fig. 2.2). This is called diffuse reflection.

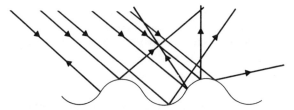

Fig. 2.2 Reflection at an irregular surface—diffuse reflection.

It is by diffuse reflection that most objects (except self-luminous ones) are seen, e.g. furniture, etc. A perfect reflecting surface (free from irregularities causing diffuse reflection) would itself be invisible. Only the image formed by light reflected in it would be seen.

Reflection at a plane surface: plane mirrors

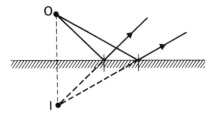

Fig. 2.3 Reflection at a plane surface—point object.

In Fig. 2.3 light from object, O, is reflected at the surface according to the laws of reflection. If the reflected rays are produced behind the surface, they all intersect at point I, the image of object, O.

The brain always assumes that an object is situated in the direction from which light enters the eye. Light from object O appears to come from point I, the image of O. However, if the observer actually goes to point I, there is no real image present—it could not be captured on a screen. Such images are called *virtual*. Images which can be captured on a screen are called *real* images.

Fig. 2.4 Reflection at a plane surface—extended object.

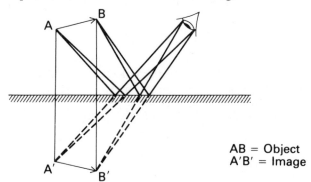

AB = Object
A'B' = Image

The image of an object formed by reflection at a plane surface has the following characteristics. It is upright (erect), virtual, and laterally inverted. It lies along a line perpendicular to the reflecting surface and is as far behind the surface as the object is in front of it.

Rotation of a plane mirror   If a plane mirror is rotated while light is incident upon its centre of rotation, the reflected ray is deviated through an angle equal to twice the angle of rotation of the mirror.

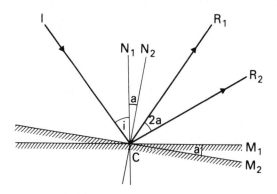

Fig. 2.5   Rotation of a plane mirror.

By the laws of reflection:
Angle of incidence   = angle of reflection
therefore,
angle between incident = angle of incidence + angle of reflection
   and reflected ray

                     = 2 × angle of incidence

With mirror at M_1, and its normal N_1,
   Angle $ICR_1 = 2i$

After rotation of the mirror to M_2, with normal N_2,
   Angle $ICR_2 = 2(i+a)$
   where a is the angle of rotation.

The angle through which the reflected ray is deviated when the mirror rotates from M_1 to M_2 is angle R_1CR_2
   But $R_1CR_2 = ICR_2 - ICR_1$
            $= 2(i+a) - 2i$
            $= 2a$.

Reflection at spherical reflecting surfaces

A reflecting surface having the form of a portion of a sphere is called a spherical mirror. If the reflecting surface lies on the inside of the curve, it is a *concave* mirror. If the reflecting surface lies on the outside of the curve, the mirror is a *convex* mirror.

Fig. 2.6 Spherical mirror— reflection of parallel light

a Concave mirror

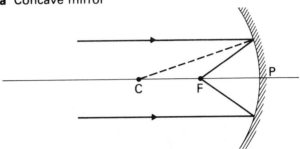

b Convex mirror

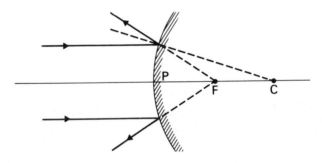

The *centre of curvature*, C, is the centre of the sphere of which the mirror is a part.

The *pole of the mirror*, P, is the centre of the reflecting surface.

$$CP = \text{the radius of curvature, r}$$

An *axis* is any line passing through the centre of curvature and striking the mirror. That passing through the pole of the mirror is called the *principal axis*; any other is a *subsidiary axis*.

Rays parallel to the principal axis are reflected towards (concave) or away from (convex) the *principal focus*, F, of the mirror (Fig. 2.6). The distance FP is the *focal length*, f, of the mirror and is equal to half the radius of curvature.

Thus, the image of an object situated on the principal axis an infinite distance away, is formed at the principal focus (Fig. 2.6). The image formed by the concave mirror is real while that formed by the convex mirror is virtual. The following diagrams show the nature and situation of images formed of objects situated at a finite distance from the mirror on the principal axis. In each case the image is constructed using two rays:

1. A ray parallel to the principal axis and reflected to or away from the principal focus.
2. A ray from the top of the object, passing through the centre of curvature and reflected back along its own path.

Fig. 2.7   Image formation by the concave mirror.

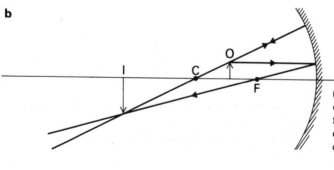

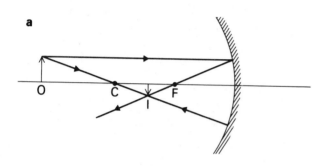

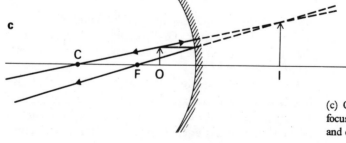

(a) Object outside the centre of curvature, C. Image real, inverted, diminished (reduced in size), lying between C, and principal focus F.

(b) Object between centre of curvature C, and principal focus F. Image real, inverted, enlarged, lying outside the centre of curvature C.

(c) Object inside principal focus, F. Image erect, virtual and enlarged.

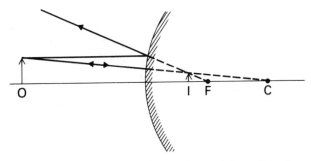

Fig. 2.8 Image formation by the convex mirror. The object may be located at any distance from the mirror. The image is virtual, erect and diminished.

For any position of the object, the position of the image formed by a spherical mirror can be calculated using the formula:

$$\frac{1}{v} - \frac{1}{u} = \frac{1}{f} = \frac{2}{r}$$

where u is the distance of the object from the mirror
v is the distance of the image from the mirror
f is the focal length of the mirror
and r is the radius of curvature of the mirror.

Also, the magnification produced by a curved mirror can be calculated. Magnification is defined as the ratio of image size to object size.

where M = magnification
i = image size
o = object size

The magnification can be calculated using the formula:

$$\frac{i}{o} = -\frac{v}{u}$$

where v is the distance of the image from the mirror
and u is the distance of the object from the mirror.

When using these formulae, the Sign Convention must be adhered to (Fig. 2.9).

Fig. 2.9 Sign convention.

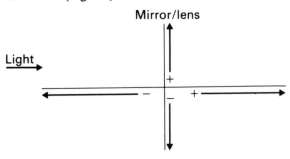

All distances are measured from the pole of the mirror (or vertex of the lens) to the point in question.

Distances measured in the same direction as the incident light are positive, those against the direction of the incident light are negative. (The incident light is usually shown coming from the left in optical diagrams.)

Image size is positive for erect images (above the principal axis) and negative for inverted images (below the principal axis).

Clinical application

The theory of curved mirrors has a major clinical application. The anterior surface of the cornea acts as a convex mirror and is used as such by the standard instruments employed to measure corneal curvature (see keratometer, Chapter 13).

Images formed by the reflecting surfaces of the eye are called Catoptric Images, and are described in Chapter 9.

Chapter 3   **Refraction of Light**

Refraction is defined as the change in direction of light when it passes from one transparent medium into another of different optical density. The incident ray, the refracted ray and the normal all lie in the same plane.

The velocity of light varies according to the density of the medium through which it travels. The more dense the medium the slower the light passes through it. When a beam of light strikes the interface separating a less dense medium from a denser one obliquely (Fig. 3.1) the edge of the beam which arrives first, A, is retarded on entering the denser medium.

Fig. 3.1 Refraction of beam of light entering an optically dense medium from air.

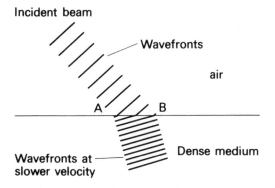

Incident beam

Wavefronts

air

A   B

Dense medium

Wavefronts at slower velocity

The opposite side of the beam, B, is meanwhile continuing at its original velocity. The beam is thus deviated as indicated in Fig. 3.1, being bent towards the normal as it enters the denser medium.

A comparison of the velocity of light in a vacuum and in another medium gives a measure of the *optical density* of that medium. This measurement is called the *absolute refractive index*, n, of the medium.

Absolute refractive index $= {}_{vacuum}n_{medium}$

$$= \frac{\text{Velocity of light in vacuum}}{\text{Velocity of light in medium}}$$

29

As the optical density of air as a medium is negligible under normal conditions,

Refractive index $= {}_{air}n_{medium}$

$$= \frac{\text{Velocity of light in air}}{\text{Velocity of light in medium}}$$

Examples of refractive index are:

Air	$= 1$
Water (incl. aqueous)	$= 1.33$
Cornea	$= 1.37$
Crystalline lens	$= 1.38\text{--}1.42$
Crown glass	$= 1.52$
Flint glass	$= 1.6$
Diamond	$= 2.5$

The absolute refractive index of any material can be determined using a refractometer.

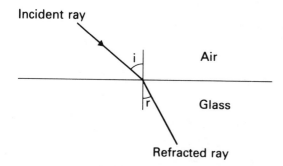

Incident ray

i   Air

r   Glass

Refracted ray

Fig. 3.2   Refraction of light entering an optically dense medium from air.

On entering an optically dense medium from a less dense medium light is deviated *towards* the normal. (The 'normal' being a line perpendicular to the interface at the point of refraction.)

The incident ray makes an angle, i, *the angle of incidence*, with the normal. The angle between the refracted ray and the normal is called the *angle of refraction*, r.

These angles are governed by the refractive indices of the media involved according to Snell's Law.

Snell's Law states that the incident ray, refracted ray and the normal all lie in the same plane and that the angles of incidence, i, and refraction, r, are related to the refractive index, n, of the media concerned by the equation

$$\text{medium}_1 n_{\text{medium}_2} = \frac{\sin i}{\sin r}$$

Where the first medium is a vacuum, n is the Absolute Refractive Index, and in air, n is the Refractive Index.

If, however, the interface is between two denser media of differing optical densities, e.g. water and glass, then the value of n for that interface may be calculated as follows

$$_{\text{water}}n_{\text{glass}} = \frac{n_{\text{glass}}}{n_{\text{water}}}$$

More generally, on passing from medium$_1$ into medium$_2$, the index of refraction is given by

$$_1 n_2 = \frac{n_2}{n_1}$$

Fig. 3.3 Refraction of light through parallel-sided slab of glass.

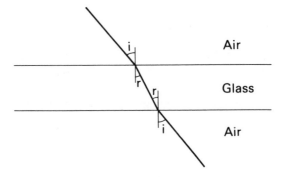

Light passing obliquely through a plate of glass (Fig. 3.3) is deviated laterally and the emerging ray is parallel to the incident ray. Thus the direction of the light is unchanged but it is laterally displaced.

It should be remembered that some reflection also occurs at every interface (Chapter 2) even though in this case most of the incident light passes onwards by refraction.

Fig. 3.4 illustrates the use of a sheet of glass as an image-splitter, e.g. the teaching mirror of the indirect ophthalmoscope. Most of the light is refracted across the glass to the examiner's eye. However, a small proportion is reflected at the anterior surface of the glass and enables an observer to see the same view as the examiner.

Observer

Examiner

Fig. 3.4   Parallel-sided glass
sheet used as an image-splitter.

Refraction of light at a curved interface

Light passing across a curved interface between two
media of different refractive indices obeys Snell's Law. A
convex spherical curved surface causes parallel light to
converge to a focus, if n_2 is greater than n_1 or to diverge
as from a point focus if n_2 is less than n_1.

Fig. 3.5   Refraction of light at
a convex refracting interface,
e.g. cornea.

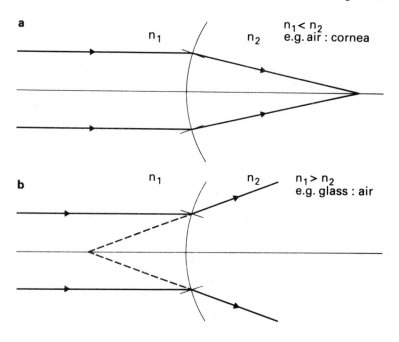

a

n_1 n_2 $n_1 < n_2$
e.g. air : cornea

b

n_1 n_2 $n_1 > n_2$
e.g. glass : air

The refracting power or vergence power of such a surface is given by the formula

$$\text{Surface power} = \frac{n_2 - n_1}{r}$$

where r is the radius of curvature of the surface in metres according to the sign convention (see p. 27) and the surface power is measured in dioptres (see p. 52).

Surface power is positive for converging surfaces and negative in sign for diverging surfaces. The anterior surface of the cornea is an example of such a refracting surface, and its power accounts for most of the refracting power of the eye.

Fig. 3.6 Real and apparent depth.

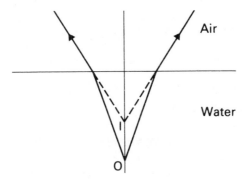

Objects situated in an optically dense medium appear displaced when viewed from a less dense medium. This is due to refraction of the emerging rays which now appear to come from a point I, the virtual image of object O. Objects in water seem less deep than they really are, e.g. one's toes in the bath.

This principle applies also to surgical instruments in the anterior chamber of the eye. For example, when making a Graefe section the knife in the anterior chamber appears to be more superficial than it really is. Therefore, to effect an exit 1 mm behind the opposite limbus, the tip is aimed at the limbus itself.

Fig. 3.7 Graefe knife in anterior chamber—real and apparent depth.

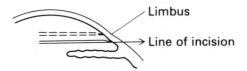

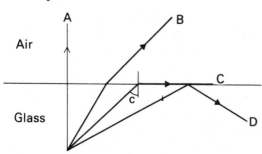

Fig. 3.8   Total internal reflection.

Rays emerging from a denser medium to a rarer medium suffer a variety of fates, depending on the angle at which they strike the interface. Ray A strikes at 90° to the interface and is undeviated. Ray B emerges after refraction. As the rays meet the interface more obliquely, a stage is reached where the refracted ray, Ray C, runs parallel with the interface. The angle c is called the *Critical Angle*. Rays striking more obliquely still fail to emerge from the denser medium and are reflected back into it as from a mirror. This is called *total internal reflection*.

This phenomenon is used in optical instruments. Prisms make excellent reflectors by total internal reflection (Chapter 4, page 45). Fibre optic cables,which are used to deliver light from a remote source to the point where it is required, also depend on total internal reflection. Examples include the surgical intraocular light source and the transmission of laser light from the laser tube to the delivery system of the laser slit lamp.

Fibre optic cables consist of many fine transparent fibres bound together in a flexible external protective sheath. Light enters the end of each fibre and is reflected onward by total internal reflection until it emerges from the far end without significant loss of radiant energy (Fig. 3.9). Bending the optical fibre does not impair its efficiency.

Total internal reflection also occurs at surfaces within the eye, notably the cornea:air interface and prevents visualization of parts of the eye, e.g. the angle of the anterior chamber and peripheral retina.

Fig. 3.9   Total internal reflection — single fibre of fibre optic cable.

Fig. 3.10 Total internal reflection at the cornea of light from the angle of the anterior chamber.

The problem is overcome by applying a contact lens made of material with a higher refractive index than the eye and filling the space between eye and lens with saline, thus destroying the cornea : air refracting surface and allowing visualization of the anterior chamber angle (gonioscopy) and peripheral retina (3-mirror).

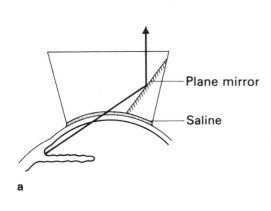

Plane mirror

Saline

a

Fig. 3.11 (a) Gonioscopy lens. (b) 3-mirror contact lens.

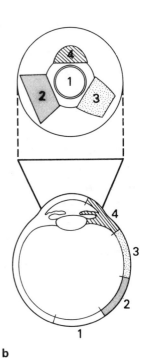

b

Dispersion of light

So far, this discussion of refraction has overlooked the fact that white light is composed of varying wavelengths. In fact, the refractive index of any medium differs slightly for light of different wavelengths.

Light of shorter wavelength is deviated more than light of longer wavelength, e.g. blue light is deviated more than red. The refractive index of a material is normally taken to mean that for the yellow sodium flame.

The angle formed between the red and blue light around the yellow (Fig. 3.12) indicates the *dispersive power* of the medium (cf chromatic aberration, p. 75). This is not related to the refractive index of the material.

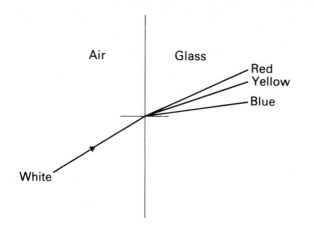

Fig. 3.12   Diagram to show dispersion of light. (The angles involved are exaggerated.)

The rainbow—total internal reflection and dispersion

Fig. 3.13   Formation of the primary rainbow. Path of light within one raindrop.

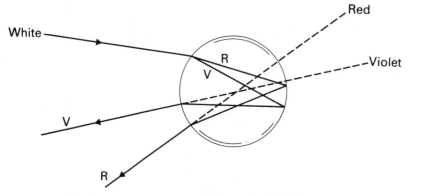

When sunlight enters a raindrop it is dispersed into its constituent spectral colours. Under certain circumstances, the angle of incidence is such that total internal reflection then occurs within the drop. The dispersed light finally emerges, each wavelength or colour making a different angle with the horizon. To see the rainbow, the observer must look away from the sun.

The observer receives only a narrow pencil of rays from each drop, i.e. only one colour. The whole rainbow is the result of rays received from a bank of drops at increasing angle to the observer's eye (see Fig. 3.14). Violet, the colour making the smallest angle to the horizon, is received from the lower drops while red, making the greatest angle with the horizon is received from the highest drops. Thus the red is on the outside of the primary rainbow.

The secondary rainbow is formed by rays that have twice undergone total internal reflection within the raindrops, and the colours are seen in reverse order—violet is on the outside of the bow.

Fig. 3.14 Formation of primary rainbow—composite diagram.

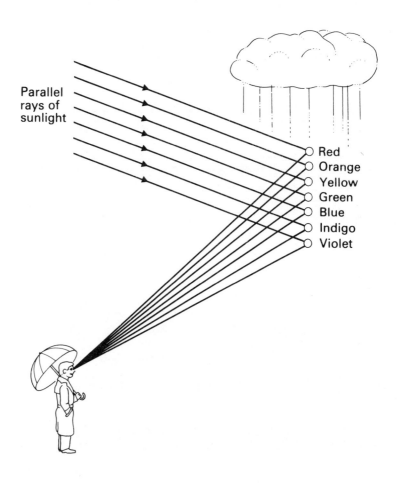

Parallel
rays of
sunlight

Red
Orange
Yellow
Green
Blue
Indigo
Violet

Prisms

Chapter 4

A prism is defined as a portion of a refracting medium bordered by two plane surfaces which are inclined at a finite angle.

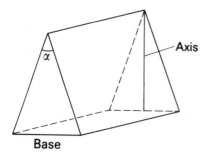

Fig. 4.1 Prism. The refracting angle (α).

The angle α between the two surfaces is called the *refracting angle* or *apical angle* of the prism. A line bisecting the angle is called the axis of the prism. The opposite surface is called the *base* of the prism. When prescribing prisms, the orientation is indicated by the position of the base, e.g. 'base-in', 'base-up'.

Light passing through a prism obeys Snell's law at each surface.

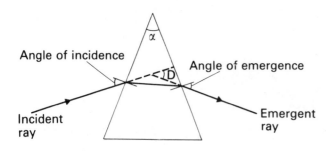

Fig. 4.2 Passage of light through a prism.

The ray is deviated towards the base of the prism. The net change in direction of the ray, angle D, is called the *angle of deviation.*

For a prism in air, the angle of deviation is determined by three factors:

1. The refractive index of the material of which the prism is made
2. The refracting angle, α, of the prism
3. The angle of incidence of the ray considered.

For any particular prism, the angle of deviation D is least when the angle of incidence equals the angle of emergence. Refraction is then said to be symmetrical and the angle is called the *angle of minimum deviation*. Under these conditions the angle of deviation is given by the formula

$$D = (n-1)\, \alpha$$
$$\text{Thus, } D = (1.5-1)\, \alpha$$
$$= \frac{\alpha}{2}$$

In other words, *the angle of deviation equals half the refracting angle of the prism.*

Fig. 4.3 Image formation by a prism.

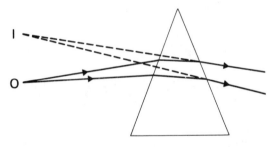

The image formed by a prism is erect, virtual and *displaced towards the apex of the prism.* Deviation is reduced to a minimum when light passes through the prism symmetrically.

There are two primary positions in which the power of a prism may be specified, the position of minimum deviation and the Prentice position. In the Prentice position one surface of the prism is normal to the ray of light so that all the deviation takes place at the other surface of the prism (Fig. 4.4).

The deviation of light in the Prentice position is greater than that in the position of minimum deviation, because in the Prentice position the angle of incidence does not equal the angle of emergence. Therefore the Prentice position power of any prism is greater than its power in the position

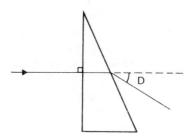

Fig. 4.4   The Prentice position of a prism.

of minimum deviation.

It is the Prentice position power which is normally specified for ophthalmic prisms, because it is convenient to mount a prism with one face approximately perpendicular to the line of sight. If a high power ophthalmic prism is not used in the Prentice position, e.g. a prism bar held in the frontal plane rather than perpendicular to the visual axis, a considerable error will result. For the same reason it is not satisfactory to stack prisms one on top of another, because the light entering the second and subsequent prisms will not be at the correct angle of incidence. The effective power of such a stack will be significantly different from the sum of the powers of the component prisms.

NOTATION OF PRISMS

The power of any prism can be expressed in various units.

The prism dioptre (Δ)

A prism of one prism dioptre power (1^Δ) produces a linear apparent displacement of 1 cm, of an object, O, situated at 1 m (Fig. 4.5).

Fig. 4.5   The prism dioptre and angle of apparent deviation (θ).

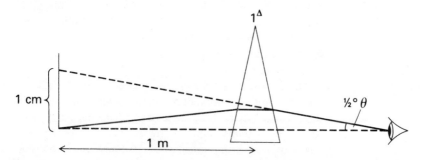

Angle of apparent deviation

The apparent displacement of the object, O, can also be measured in terms of the angle θ, the angle of apparent deviation (Fig. 4.5). Under conditions of ophthalmic usage a prism of 1 prism dioptre power produces an angle of apparent deviation of $1/2°$

Thus 1 prism dioptre $= 1/2°$.

The centrad (∇)

This unit differs from the prism dioptre only in that the image displacement is measured along an arc 1 m from the prism.

Fig. 4.6 The centrad (∇).

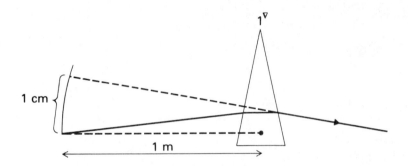

1 cm

1^{∇}

1 m

The centrad produces a very slightly greater angle of deviation than the prism dioptre, but the difference, in practice, is negligible.

Refracting angle

A prism may also be described by its refracting angle (Fig. 4.1). However, unless the refractive index of the prism material is also known, the prism power cannot be deduced.

SUMMARY OF PRISM UNITS

Thus a glass prism of refracting angle 10° (a ten degree prism) deviates light through 5° and has a power of 10 prism dioptres ($10^{\triangle}$), assuming its refractive index is 1.5.

INTERPRETATION OF ORTHOPTIC REPORTS

Armed with the knowledge that 1 prism dioptre = $\frac{1}{2}°$, orthoptic reports become intelligible to the clinician.

The orthoptist measures the angle of squint by two methods. Using the synoptophore, she measures the angle between the visual axes of the eyes in degrees, + *signifying esotropia and* – *signifying exotropia*. Her report reads

Synopt. s̄ gls +20°

She also measures the angle of squint by the Prism Cover Test (PCT). The alternating cover test is performed, placing prisms of increasing strength before one eye, until movement of the eyes is eliminated (cf Fig. 4.7). The result is expressed in prism dioptres, *eso signifying esotropia and exo signifying exotropia*. Her report reads

PCT = distance eso +40[△]

These two statements express the same angle of squint.

USE OF PRISMS

Diagnostic

1. Assessment of squint and heterophoria.

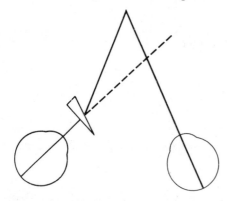

Fig. 4.7   Use of a prism in squint.

(a) Measurement of angle objectively by prism cover test.
(b) Measurement of angle subjectively by Maddox rod.
(c) To assess likelihood of diplopia after proposed squint surgery in adults.

(d) Measurement of fusional reserve. Increasingly powerful prisms are placed before one eye until fusion breaks down. This is very useful in assessing the presence of binocular single vision in children under 2 years old.

(e) The four dioptre prism test. This is a delicate test for small degrees of esotropia (microtropia). A four dioptre prism placed base-out before the deviating eye causes no movement as the image remains within the suppression scotoma. When placed before the normal (fixing) eye, movement occurs.

2. Assessment of simulated blindness. If a prism is placed in front of a seeing eye, the eye will move to regain fixation.

Forms of prisms used in diagnosis

Forms of prism used in assessment include single unmounted prisms, the prisms from the trial lens set and prism bars. These are bars composed of adjacent prisms of increasing power.

Therapeutic

1. Convergence insufficiency. The commonest therapeutic use of prisms in the orthoptic department is in building up the fusional reserve of patients with convergence insufficiency. The prisms are used base-out during the patients' exercise periods. *They are not worn constantly*.

2. To relieve diplopia in certain cases of squint. These include decompensated heterophorias, small vertical squints and some paralytic squints with diplopia in the primary position. Prisms are reserved for those patients for whom surgery is not indicated.

Forms of therapeutic prisms

(a) Temporary wear. Forms of prisms used in treatment include clip-on spectacle prisms for trial wear. An improvement on these are *Fresnel prisms*, which are available in all powers employed clinically. A Fresnel prism consists of a plastic sheet of parallel tiny prisms of identical refracting angle. The overall prismatic effect is the same as a single large prism. The sheets are lighter than a glass prism and can be stuck onto the patient's glasses.

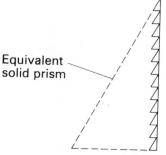

Fig. 4.8   Fresnel prism.

(b) Permanent wear. Permanent incorporation of a prism into a patient's spectacles can be achieved by decentring the spherical lens already present, see p. 56. Alternatively, prisms can be mounted in spectacles.

Notes on prescription of prisms

Generally, when prescribing prisms, the correction is split between the two eyes.

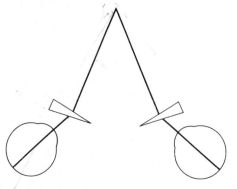

Fig. 4.9   Convergence with prismatic correction.

To correct convergence the prisms must be base-out, e.g. $8^\triangle$ base-out R and L.

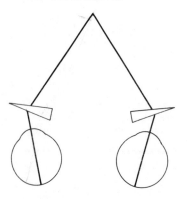

Fig. 4.10   Divergence with prismatic correction.

To correct divergence the prisms must be base-in, e.g. 6^Δ base-in R and L.

Fig. 4.11 Vertical deviation with prismatic correction.

Hypertropic eye

Hypotropic eye

To correct vertical deviation the orientation of the prisms is opposite for the two eyes, e.g.

2^Δ base-down RE
2^Δ base-up LE for R hypertropia

The apex of the prism must always be placed towards the direction of deviation of the eye.

Prisms in optical instruments

Right angle prism

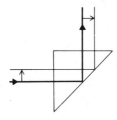

Deviation 90°

Porro prism

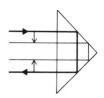

Deviation 180°
Image inverted but not
transposed left to right

Fig. 4.12 Some prisms used in optical instruments.

Dove prism

No deviation
Image inverted but not
laterally transposed

Prisms are commonly used in ophthalmic instruments as reflectors of light. The prism is designed and orientated so that total internal reflection occurs within it.

It can be seen that prisms give greater flexibility in dealing with an image than do mirrors. There are many possible systems available.

Instruments in which prisms are used include the slit lamp microscope, the applanation tonometer and the keratometer (see Chapter 13).

Chapter 5   Spherical Lenses

A lens is defined as a portion of a refracting medium bordered by two curved surfaces which have a common axis. When each surface forms part of a sphere, the lens is called a spherical lens. Various forms of spherical lens are possible (Fig. 5.1), some having one plane surface. This is acceptable because a plane surface can be thought of as part of a sphere of infinite radius.

Fig. 5.1 Basic forms of spherical lenses.

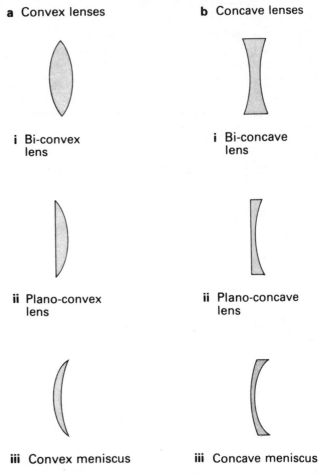

a Convex lenses

b Concave lenses

i Bi-convex lens

i Bi-concave lens

ii Plano-convex lens

ii Plano-concave lens

iii Convex meniscus

iii Concave meniscus

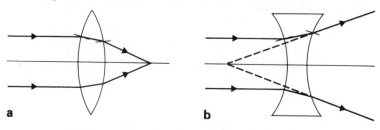

A convex lens causes convergence of incident light, whereas a concave lens causes divergence of incident light.

The total vergence power of a spherical lens depends on the vergence power of each surface (see Chapter 3, p. 32) and the thickness of the lens. Most of the lenses used in ophthalmology are thin lenses, and for a thin lens the thickness factor may be ignored. Thus the total power of a thin lens is the sum of the two surface powers. Refraction can be thought of as occurring at the principal plane of the lens and in the following lens diagrams only the principal plane is shown. Refraction by thick lenses is more complicated, and the theory of the thick lens is dealt with in Chapter 9 as it is more relevant to the study of the refracting mechanism of the eye.

Fig. 5.2 Light passing through a lens obeys Snell's Law at each surface. (a) Convex lens (b) Concave lens.

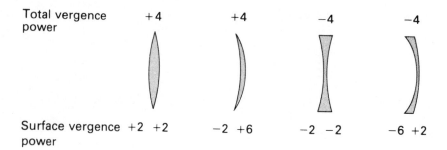

Fig. 5.3 Vergence power of thin spherical lenses.

Fig. 5.4 Cardinal points of thin spherical lenses. (a) Convex (b) Concave.

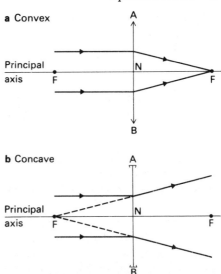

In Fig. 5.4 a and b the *principal plane* of the lens is shown, AB. (Note that in ray diagrams the convex or concave nature of a thin lens is shown by the appropriate symbol at each end of the line that indicates the principal plane.) The point at which the principal plane and principal axis intersect is called the *principal point* or *nodal point*, N, of the lens. Rays of light passing through the nodal point are undeviated.

Light parallel to the principal axis is converged to or diverged from the point F, the *principal focus*. As the medium on both sides of the lens is the same (air) parallel light incident on the lens from the opposite direction, i.e. from the right in Fig. 5.4 will be refracted in an identical way. There is therefore a principal focus on each side of the lens, equidistant from the nodal point. The two principal foci are by convention distinguished from each other according to the following rules. (It must be remembered that in optical diagrams light is always shown travelling from left to right.)

Fig. 5.5 The principal foci of thin spherical lenses. (See Fig. 5.5b on next page)

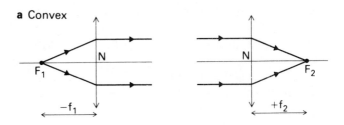

b Concave

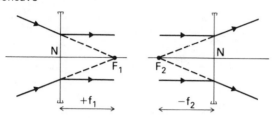

The *first principal focus*, F_1, is the point of origin of rays which after refraction by the lens, are parallel to the principal axis. The distance $F_1 N$ is the first focal length f_1.

Incident light parallel to the principal axis is converged to or diverged from the *second principal focus*, F_2. The distance F_2N is the second focal length, f_2. By the sign convention (see p. 27) f_2 has a positive sign for the convex lens, and a negative sign for the concave lens.

Lenses are designated by their second focal length. Thus, convex or converging lenses are sometimes called 'plus lenses', and are marked with a +, while concave or diverging lenses are known as 'minus lenses' and are marked with a −.

If the medium on either side of the lens is the same, e.g. air, then $f_1 = f_2$. However, if the second medium differs from the first, e.g. as in the case of a contact lens, then f_1 will not equal f_2 (cf. refraction at curved interfaces Chapter 3).

Thin lens formula

As with spherical mirrors, the position and nature of the image formed by a spherical lens depends on the position of the object and a similar formula applies:

$$\frac{1}{v} - \frac{1}{u} = \frac{1}{f_2}$$

where v is the distance of the image from the principal point

u is the distance of the object from the principal point

and f_2 is the second focal length.

For an object in any position, the image can be constructed using two rays:

1. A ray from the top of the object which passes through the principal point undeviated
2. A ray parallel to the principal axis, which after refraction passes through (convex) or away from (concave) the second principal focus.

Fig. 5.6 Image formation by a thin convex lens.

a Object outside F_1. Image real, inverted, outside F_2

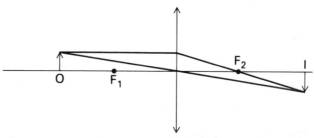

b Object at F_1. Image virtual, erect, at infinity

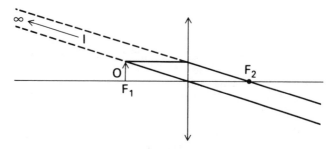

c Object inside F_1. Image virtual, erect, magnified, Further from lens than object

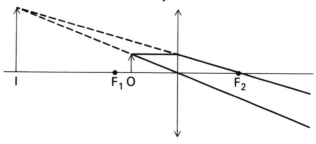

Real object at any position produces virtual, erect, diminished image, inside F_2.

Fig. 5.7 Image formation by a thin concave lens.

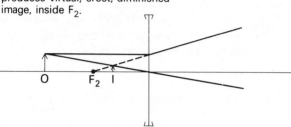

Dioptric power of lenses. Vergence

Lenses of shorter focal length are more powerful than lenses of longer focal length. Therefore the unit of lens power, the *dioptre*, is based on the reciprocal of the second focal length. The reciprocal of the second focal length expressed in metres, gives the *vergence power* of the lens in dioptres (D) thus:

$$F = \frac{1}{f_2}$$

where F is the vergence power of the lens in dioptres and f_2 is the second focal length *in metres*.

A converging lens of second focal length +5 cm has a power of

$$+\frac{1}{0.05} \text{ or } + 20 \text{ D.}$$

Likewise, a diverging lens of second focal length −25 cm has a power of

$$-\frac{1}{0.25} \text{ or } -4 \text{ D.}$$

It is possible to think of the other terms in the lens equation in a similar way. The reciprocal of object and image distances *in metres* gives a dioptric value which is

Fig. 5.8 Vergence of rays.
(a) Vergence at the lens
$$= \frac{1}{0.25}$$
$$= 4 \text{ Dioptres}$$
(b) Vergence at the lens
$$= \frac{1}{0.10}$$
$$= 10 \text{ Dioptres}$$

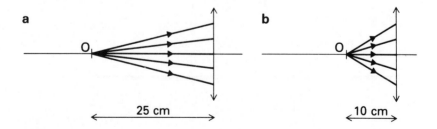

a   b

25 cm   10 cm

a measure of the vergence of the rays between object or image and the lens. In other words, it is a measure of the degree of convergence or divergence of the rays in question.

The concept of vergence is an important aid to the understanding of the optics of accommodation and presbyopia.

Magnification formulae

Linear magnification The linear magnification produced by a spherical lens can be calculated from the basic formula:

$$\text{Linear magnification} = \frac{I}{O} = \frac{v}{u}$$

where I is the image size

O is the object size

v is the distance of the image from the principal plane

u is the distance of the object from the principal plane

Fig. 5.9 Linear magnification. Convex lens. Object within focal length.

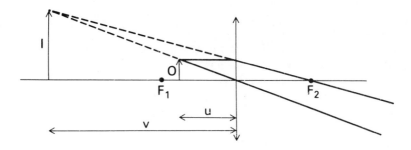

Angular magnification

In ophthalmic practice, actual image and object size is of less importance than the angle subtended at the eye, because the angle subtended governs the retinal image size.

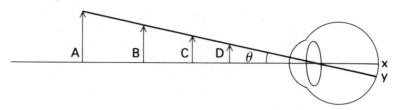

Fig. 5.10a shows that objects A, B, C and D all subtend angle θ at the eye and produce a retinal image xy. They are all therefore of identical *apparent size*. Apparent size is given by the ratio of object (or image) size divided by its distance from the eye, which is, of course, tan θ. (See Appendix).

Fig. 5.10a Apparent size—visual angle.

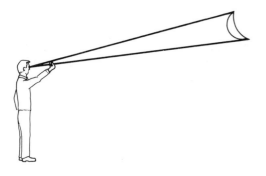

Fig. 5.10b Apparent size—visual angle illustrated. A coin held at arms-length obscures the view of the moon.

When considering the eye, the angles encountered are small. For small angles the value of tan θ can be taken to be equal to the angles themselves.

The concept of apparent size permits the assignment of a definite magnitude to an image at infinity, such as that formed by a convex lens when the object is situated at the first principal focus.

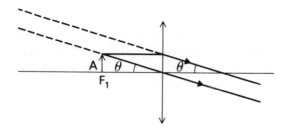

Fig. 5.11 Image formation by convex lens with object at first principal focus—the Magnifying Glass or Loupe.

The object and its infinitely distant image subtend the same angle, θ, at the lens (Fig. 5.11) and also at the eye, if the eye is brought very close to the lens. The angular magnification is therefore unity, i.e. apparent object size and apparent image size are the same.

However, the use of a convex lens enables the eye to view the object at a much shorter distance than would be possible unaided, and to retain a distinct image. As the object approaches the eye it subtends a greater angle at the eye and the retinal image size increases.

The simple magnifying glass (the Loupe)

Fig. 5.12 The simple magnifying glass (the Loupe) (a) Object viewed at near point of unaided eye, 25 cm, subtends angle θ_1 at the eye.

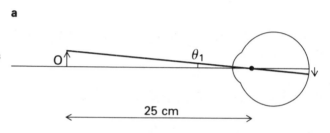

(b) Object viewed close to the eye through a convex lens, with object at first principal focus of convex lens. Object and image subtend angle θ_2 at the eye.

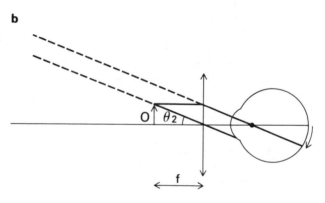

The magnifying power of the lens under these conditions can be expressed as follows:

$$\text{Magnifying Power} = \frac{\text{Apparent Size of Image}}{\text{Apparent Size of Object}} \text{ at 25 cm from the eye}$$

alternatively

$$\text{Magnifying Power, M,} = \frac{\tan \theta_2}{\tan \theta_1}$$

But

$$\tan \theta_1 = \frac{o}{25}$$

and

$$\tan \theta_2 = \frac{o}{f}$$

thus

$$M = \frac{o}{f} \times \frac{25}{o}$$

$$= \frac{25}{f}$$

But 25 cm $=\frac{1}{4}$ m, and $\frac{1}{f} = F$ dioptres, where F is the power of the lens in dioptres.

Therefore $M = \frac{F}{4}$

Thus, the commonly used '×8' loupe has a lens power of +32 dioptres.

Spherical lens decentration and prism power

Rays of light incident upon a lens outside its axial zone are deviated towards (convex lens) or away from (concave lens) the axis. Thus the peripheral portion of the lens acts as a prism.

The refracting angle between the lens surfaces grows larger as the edge of the lens is approached (Fig. 5.13). Thus the prismatic effect increases towards the periphery of the lens.

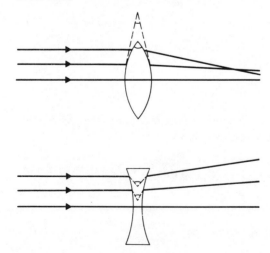

Fig. 5.13 Prismatic deviation by spherical lenses.

Use of a non-axial portion of a lens to gain a prismatic effect is called *decentration* of the lens. Lens decentration is frequently employed in spectacles where a prism is to be incorporated. On the other hand, poor centration of spectacle lenses, especially high power lenses, may produce an unwanted prismatic effect. This is a frequent cause of spectacle intolerance especially in patients with aphakia or high myopia.

It is thus of importance to be able to predict the prismatic power gained by decentring a spherical lens. This is given by the formula

$$P = F \times D$$

where   P is the prismatic power in prism dioptres

           F is the lens power in dioptres

and   D is the decentration *in centimetres*

The increasing prismatic power of the more peripheral parts of a spherical lens is the underlying mechanism of spherical aberration (see p. 77). Furthermore, it causes the troublesome ring scotoma and jack-in-the-box effect which give rise to great difficulty to those wearing high power spectacle lenses (p. 113).

Astigmatic Lenses   Chapter 6

All the meridians of each surface of a spherical lens have the same curvature (as parts of a sphere), and refraction is symmetrical about the principal axis.

In an astigmatic lens, all meridians do not have the same curvature, and a point image of a point object cannot be formed. There are two types of astigmatic lenses, namely cylindrical and toric lenses.

CYLINDRICAL LENSES

These lenses have one plane surface and the other forms part of a cylinder.

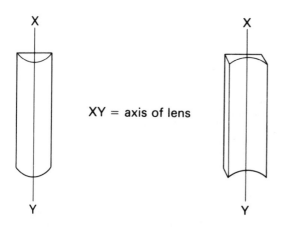

XY = axis of lens

Fig. 6.1 Cylindrical lens.

Thus, in one meridian the lens has no vergence power and this is called the *axis of the cylinder*. In the meridian at right angles to the axis, the cylinder acts as a spherical lens. The total effect is the formation of a line image of a point object. This is called the *focal line*. It is parallel to the axis of the cylinder (Fig. 6.2).

Fig. 6.2 Image formation by convex cylindrical lens of point object, O.

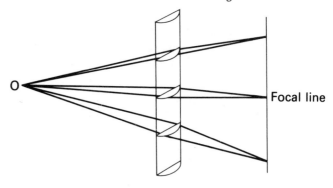

THE MADDOX ROD

This useful device, used in the diagnosis of extraocular muscle imbalance, consists of a series of powerful convex cylindrical lenses mounted side by side in a trial lens.

Fig. 6.3 The Maddox rod.

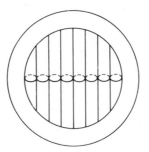

The patient views a distant white point source of light through the Maddox rod, which is placed close to the eye (in the trial frame). The spot light must be far enough away for its rays to be parallel on reaching the patient (at least 6 m).

Fig. 6.4 Optics of Maddox rod. Light in the meridian parallel to axis of the rod.

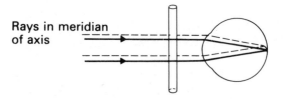

Rays in meridian of axis

Light in the meridian parallel to the axis of each cylinder passes through undeviated and *is brought to a focus by the eye* (Fig. 6.4). The Maddox rod consists of a row of such cylinders, and thus a row of foci are formed on the retina (Fig. 6.5). These foci join up and are seen as a line of light which lies *at 90° to the axis of the Maddox rod.*

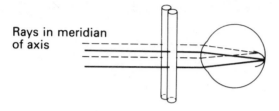

Rays in meridian
of axis

Fig. 6.5 Optics of Maddox rod. To show formation of line of foci by adjacent elements of Maddox rod.

Meanwhile, light incident on the Maddox rod in the meridian at 90° to its axis is converged by each cylinder to a real line focus between the rod and the eye (Fig. 6.6). This focus is too close to the eye for a distinct image to be formed on the retina by the focusing mechanism of the eye. This light is therefore scattered over a wide area of retina (Fig. 6.6) and does not confuse the perception of the composite line image described above (Fig. 6.5).

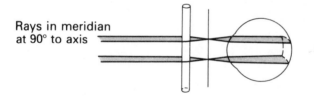

Rays in meridian
at 90° to axis

Fig. 6.6 Optics of Maddox rod. Light incident in the meridian at 90° to axis of Maddox rod.

Remember that the line seen by the patient lies at 90° to the axis of the Maddox rod and is formed by the focusing mechanism of the eye. It is not the real line image of the Maddox rod.

The glass of the Maddox rod is tinted red so the composite line image seen by the patient is also red.

Use of the Maddox rod to test muscle balance

To test muscle balance the Maddox rod is placed close in front of the right eye (in the trial frame) and the distant white spot light is viewed with both eyes. The right eye, therefore, sees a red line, at 90° to the axis of the Maddox rod, while the left eye sees the white spot light.

Fig. 6.7 The patient's view (Maddox rod before the right eye).

Thus the two eyes see dissimilar images and are dissociated, allowing any muscle imbalance to become manifest. To test for horizontal imbalance, the rod must be horizontal to give a vertical line and vice versa.

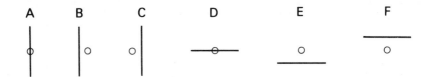

Remember that the eye behind the Maddox rod (conventionally the right) is deviating in the opposite direction to that indicated by the red line.

A   Horizontal orthophoria
B   Exophoria   The patient has crossed diplopia and this indicates an exophoria ('X'ed diplopia ≡ e<u>x</u>o)
C   Esophoria
D   Vertical orthophoria
E   Right hyperphoria
F   Right hypophoria (= Left hyperphoria)

Any deviation is measured by placing prisms before the left eye until the orthophoric situation is achieved.

TORIC SURFACE. TORIC LENSES

Imagine that the cylindrical lens in Fig. 6.1 is picked up by its ends and bent so that the axis XY becomes an arc of a circle. The previously cylindrical surface is now curved in both its vertical and horizontal meridians, but not to the same extent. It is now called a *toric* surface. The meridians of maximum and minimum curvature are called the *principal meridians* and in ophthalmic lenses these are at 90° to each other.

Dotted radii = vertical meridian
Shaded sector = horizontal meridian

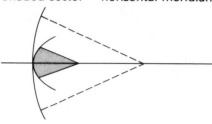

Fig. 6.8   Toric surface. Principal meridians with radii and centres of curvature.

The principal meridian of minimum curvature, and therefore minimum power, is called the *base curve*.

Toric lenses

Lenses with one toric surface are known as toric lenses, or sphero-cylindrical lenses. Such lenses do not produce a single defined image because the principal meridians form separate line foci at right angles to each other.

Between the two line foci the rays of light form a figure known as Sturm's conoid (after the mathematician Sturm who described it in 1838).

Fig. 6.9   Image formed by toric astigmatic lens—Sturm's conoid.

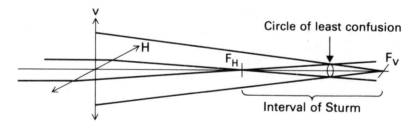

The distance between the two line foci is called the interval of Sturm. The plane where the two pencils of light intersect is called the circle of least confusion or the circle of least diffusion. Blur circle images only are formed at all other planes lying between F_H and F_V.

A toric lens can be thought of as a spherical lens with a cylindrical lens superimposed upon it. Toric lenses may be defined numerically as a fraction, the spherical power being the numerator and the cylindrical power the denominator. For example a toric lens with a power of +2 D in one principal meridian and +4 D in the other principal meridian can be regarded as a +2 D sphere

with a +2 D cylinder superimposed. This is therefore written as +2.0 DS
+2.0 DC

THE CROSS-CYLINDER

In clinical refraction the orientation of the trial cylinder can be checked by superimposing another cylinder with its axis lying obliquely to the axis of the trial cylinder. The power of a cylinder can be checked by superimposing further cylinders of varying power and sign in the same axis as the trial cylinder. These considerations have led to the evolution of the cross-cylinder.

The cross-cylinder is a type of toric lens used during refraction. Its use was popularized by Edward Jackson (1893–1929) and it is often referred to as 'Jackson's cross-cylinder'.

The cross-cylinder is a sphero-cylindrical lens in which the power of the cylinder is twice the power of the sphere and of the opposite sign (Fig. 6.10a). The net result is thus the same as superimposing two cylindrical lenses of equal power but opposite sign with their axes at right angles. The lens is mounted on a handle which is placed at 45° to the axes of the cylinders.

Fig. 6.10a A cross-cylinder.
−0.50 DS
+1.0 DC

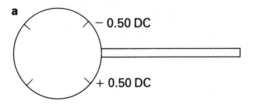

The axes marked on the lens are the axes of no power of the individual cylinders. The power of each cylinder lies at 90° to the marked axis and coincides with the marked axis (of no power) of the other cylinder (of opposite sign) (Fig. 6.10b).

Fig. 6.10b The cross-cylinder showing axes as marked on the lens and refractive power in the principal meridians.

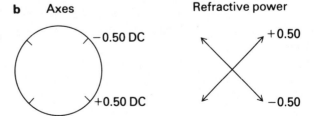

Cross-cylinders are named by the power of the cylinder, and this is marked on the handle. The cross-cylinder illustrated in Figs 6.10a & b would be designated a 1.00 Dioptre cross-cylinder. Cross-cylinders are available in two powers, 0.50 and 1.00 Dioptre. The 1.00D cross-cylinder is used to check the axis of the trial cylinder, and the power in patients with poor visual acuity. The 0.50D cross-cylinder is used to check the power of the trial cylinder where the patient has good vision.

Clinically the cross-cylinder is used to check the axis of the cylinder prescribed and then its power. It can also be used to verify that no cylindrical correction is necessary for the patient if no cylinder was detected on retinoscopy.

In practice the patient is asked to look at the line of test type two lines above the smallest he can see. This is because the cross-cylinder blurs the vision and larger letters are used to make discrimination between the positions of the cross-cylinder easier for the patient.

To check the axis, the cross-cylinder is held before the eye with its handle in line with the axis of the trial cylinder. The cross-cylinder is turned over and the patient asked which position gives a better visual result. The cross-cylinder is held in the preferred position and the axis of the trial cylinder rotated slightly towards the axis of the same sign on the cross-cylinder. The process is repeated until the trial cylinder is in the correct axis for the eye, at which time rotation of the cross-cylinder will offer equally unacceptable visual alterations to the patient.

To check the power of the trial cylinder the cross-cylinder is held with first one axis and then the other overlying the trial cylinder. This has the effect of increasing and then decreasing the power of the trial cylinder.

To confirm the absence of a cylinder, the cross-cylinder is offered as an addition to the trial sphere in four different orientations, with its + axis at 90°, 180°, 45°, and 135°. If the patient prefers one of these options to the sphere alone, a cylindrical correction is necessary. The exact axis and power can then be determined by the methods described above.

To achieve the best results from the test it is important that the patient has the clearest vision possible before the cross-cylinder is used.

Chapter 7 Optical Prescriptions Spectacle Lenses

The lenses described in Chapters 5 and 6 have many uses in ophthalmology. Lenses are used as optical aids for patients with refractive errors, in the form of spectacles or contact lenses, and as low vision aids (Chapter 12). Lenses are also an essential component of most of the instruments used in ophthalmology (Chapter 13). In this chapter the use of lenses in spectacles is discussed.

Prescription of lenses

When prescribing a spectacle lens, the properties of the lens required are specified in the following way.

A spherical lens alone is written as, for example, +2.00 DS (Dioptre Sphere) or −3.25 DS.

In the case of a cylindrical lens alone both the dioptric power and the orientation of the axis must be specified. The axis of the cylinder is marked on each trial lens by a line and trial frames are marked according to a standard international convention (Fig. 7.1).

Thus, a cylinder of −2.0 dioptre power, placed with its axis (of no power) vertical is written as −2.0 DC axis 90° (DC = Dioptre Cylinder).

Often the correction of a refractive error entails the prescription of both a spherical and a cylindrical component, i.e. a toric astigmatic correction. In such a case, at the end of refraction the trial frame contains a spherical lens (e.g. +2.0 DS) and a cylindrical lens (e.g. +1.0 DC axis 90°). The cylindrical lens is usually placed in front of the spherical lens to allow the axis line to be seen.

The prescription is written as $\dfrac{+2.00\ DS}{+1.00\ DC\ axis\ 90°}$

and this may be abbreviated to $\dfrac{+2.00}{+1.00}{\scriptstyle\downarrow 90°}$

It is useful to include an arrow to indicate the axis of the cylinder in addition to the figure to safeguard against illegible numerals.

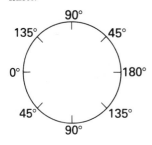

Fig. 7.1 Conventional orientation for cylindrical lenses.

Transposition of lenses

When a lens prescription is changed from one lens form to another optically equivalent form, the process is called transposition of the lens.

Simple transposition of spheres   This applies to the alteration of the lens form of spherical lenses. The lens power is given by the algebraic sum of the surface powers (Fig. 7.2).

+1.5 +1.5   0 +3.0 −1.50 +4.50 −6.00 +9.00

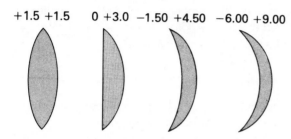

Fig. 7.2   Simple transposition of a +3.0 D spherical lens.

Simple transposition of cylinders   This is a change in the description of a toric astigmatic lens so that the cylinder is expressed in the opposite power. Simple transposition of the cylinder is often necessary when the examiner wishes to compare the present refraction with a previous prescription.

Consider the following example (Fig. 7.3).

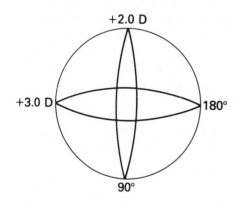

Fig. 7.3   Diagram representing the toric lens
+2.00 DS
+1.00 DC axis 90°
showing dioptric power of principal meridia.

The lens depicted in Fig. 7.3 can be described in two ways.

1. Let the cylindrical element be at axis 90°—the lens is
   now $\dfrac{+2.0 \text{ DS}}{+1.0 \text{ DC axis } 90°}$

2. Let the cylindrical element be of opposite power and
   at axis 180°—the lens is now $\dfrac{+3.0 \text{ DS}}{-1.0 \text{ DC axis } 180°}$

This change in the description of the lens may be easily
accomplished for any lens by performing the following
steps.

i. Algebraic addition of sphere and cylinder gives new
   power of sphere.
ii. Change sign of cylinder, retaining numerical power.
iii. Rotate axis of cylinder through 90° (Add 90° if the
   original axis is at or less than 90°. Subtract 90° from
   any axis figure greater than 90°).

Examples: (It is suggested that the reader cover one side of
the page to conceal the answers and work
through these examples before checking the
results).

$$\dfrac{+4.0 \text{ DS}}{+1.50 \text{ DC axis } 90°} \equiv \dfrac{+5.50 \text{ DS}}{-1.50 \text{ DC axis } 180°}$$

$$\dfrac{-2.0 \text{ DS}}{-1.0 \text{ DC axis } 170°} \equiv \dfrac{-3.0 \text{ DS}}{+1.0 \text{ DC axis } 80°}$$

$$\dfrac{+1.0 \text{ DS}}{-3.0 \text{ DC axis } 180°} \equiv \dfrac{-2.0 \text{ DS}}{+3.0 \text{ DC axis } 90°}$$

$$\dfrac{-1.50 \text{ DS}}{+2.50 \text{ DC axis } 20°} \equiv \dfrac{+1.0 \text{ DS}}{-2.50 \text{ DC axis } 110°}$$

Toric transposition   Toric transposition carries the process
one step further and enables a toric astigmatic lens to be
exactly defined in terms of its surface powers. A toric astigmatic lens is made with one spherical surface, and one toric
surface, (the latter contributing the cylindrical power).
The principal meridian of weaker power of the toric
surface is known as the *base curve* of the lens. The base
curve must be specified if toric transposition of a lens prescription is required.

The toric formula is written in two lines, as a fraction.
The top line (numerator) specifies the surface power of the
spherical surface. The bottom line (denominator) defines
the surface power and axis of the base curve, followed by
the surface power and axis of the other principal meridian
of the toric surface,

e.g.
$$\frac{+9.0 \text{ DS}}{-6.0 \text{ DC axis } 90°/-8.0 \text{ DC axis } 180°}$$

The steps of toric transposition are now defined taking the following case as an example.

'Transpose $\dfrac{+3.0 \text{ DS}}{+1.0 \text{ DC axis } 90°}$

to a toric formula to the base curve —6D'

1. Transpose the prescription so that the cylinder and the base curve are of the same sign, e.g.

   (a) $\dfrac{+3.0 \text{ DS}}{+1.0 \text{ DC axis } 90°}$

   becomes

   (b) $\dfrac{+4.0 \text{ DS}}{-1.0 \text{ DC axis } 180°.}$

2. Calculate the required power of the spherical surface (the numerator of the final formula). This is obtained by subtracting the base curve power from the spherical power given in (b) in step 1.

   e.g. $+4 \text{ D} - (-6 \text{ D}) = +10 \text{ D.}$

   Put another way, to obtain an overall power of $+4.0$ D where one surface of the lens has the power -6 D, the other surface must have the power +10 D (C.f. simple transposition of spheres).

3. Specify the axis of the base curve. As this is the weaker principal meridian of the toric surface, its axis is at 90° to the axis of the required cylinder found in (b) in step 1.

   i.e. -6 D axis 90°

4. Add the required cylinder to the base curve power with its axis as in (b) in step 1

   $-6 \text{ D} + (-1 \text{ D}) = -7 \text{ DC axis } 180°$

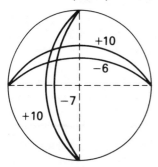

Fig. 7.4 Diagram representing toric astigmatic lens
$\dfrac{+4.0 \text{ DS}}{-1.0 \text{ DC}}$ axis 180°
with base curve −6 D.

The complete toric formula is thus

$$\frac{+10\text{ DS}}{-6\text{ DC axis }90°/-7\text{ DC axis }180°}$$

Some further examples for calculation are given below:

A. Transpose $\dfrac{+4.0\text{ DS}}{-2.0\text{ DC axis }180°} \equiv \dfrac{-4.0\text{ DS}}{+6.0\text{ DC axis }180°/+8.0\text{ DC axis }90°}$
to the base curve $+6$ D

B. Transpose $\dfrac{-2.0\text{ DS}}{+3.0\text{ DC axis }90°} \equiv \dfrac{+7.0\text{ DS}}{-6.0\text{ DC axis }90°/-9.0\text{ DC axis }180°}$
to the base curve -6 D

Identification of lenses

In clinical practice it is frequently necessary for the practitioner to identify the type and power of the patient's existing spectacles. This may be done by the following means:

Detection of lens type It is possible to determine whether a given lens is spherical, astigmatic or a prism by studying the image formed when two lines, crossed at 90°, are viewed through the lens.

Spherical lenses cause no distortion of the cross. However, when the lens is moved from side to side and up and down along the arms of the cross, the cross also appears to move. In the case of a convex lens, the cross appears to move in the opposite direction to the lens, termed as 'against movement', while a movement in the same direction as the lens, a 'with movement' is observed if the lens is concave.

Fig. 7.5 Detection of spherical lenses. (See Fig. 7.5b on next page)

a Convex = 'against movement'

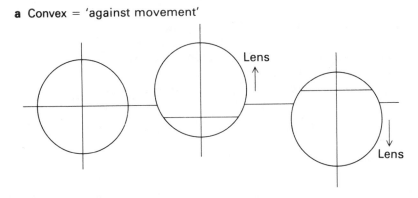

b Concave = 'with movement'

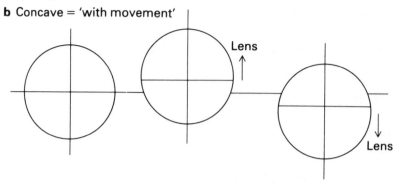

Astigmatic lenses cause distortion of the cross unless their axes coincide with the cross lines. Rotation of the lens thus causes a 'scissors' movement as the crossed lines are progressively displaced (Fig. 7.6). Rotation of a spherical lens has no effect upon the image of the crossed lines.

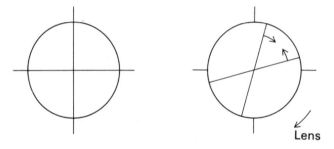

Fig. 7.6 Detection of astigmatic lenses.

Once the principal meridians of an astigmatic lens have been identified, and aligned with the cross, each meridian may be examined as for a spherical lens.

The optical centre of a lens may be found by moving the lens until one cross line is undisplaced. A line is then drawn on the lens surface, superimposed on the undisplaced cross line. The process is then repeated for the cross line at 90°. The point where the lines drawn on the lens intersect is the optical centre of the lens.

A prism has no optical centre and thus displaces one line of the cross regardless of its position with respect to the cross. Furthermore, the direction of displacement is constant.

Fig. 7.7 Detection of a prism.

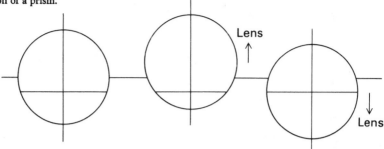

This test is most effective if the cross lines are placed at the furthest convenient distance and the lens held well away from the eye.

Neutralization of power The power of a lens can also be found using the technique described above. Once the nature of the unknown lens is determined, lenses of opposite type and known power are superimposed upon the unknown lens until a combination is found which gives no movement of the image of the cross lines when the test is performed. At this point the two lenses are said to 'neutralize' each other, and the dioptric power of the unknown lens must equal that of the trial lens of opposite sign (a +2.0 D lens neutralizes a −2.0 D lens).

In the case of astigmatic lenses, each meridian must be neutralized separately.

Spectacle lenses are named by their back vertex power (see Chapter 9). To measure this accurately, the neutralizing lens must be placed in contact with the back surface of the spectacle lens. However with many highly curved lenses this is not possible and an air space intervenes. It is then better to place the neutralizing lens against the front surface of the spectacle lens. Neutralization is thus somewhat inaccurate for curved lenses of more than about 2 dioptres power and an error of up to 0.50 dioptre may be incurred with powerful lenses. Nevertheless, for relatively low power lenses neutralization is still a very useful technique.

Lens measure The Geneva lens measure can be used to find the surface powers of a lens by measuring the surface curvature. The total power of a thin lens equals the sum of its surface powers. However, the instrument is calibrated for lenses made of crown glass (refractive index 1.523) and a correction factor must be applied in the case of lenses made of materials of different refractive indices.

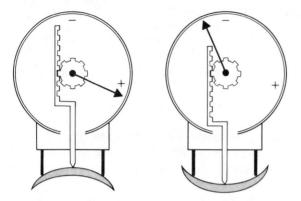

Fig. 7.8   Principle of the Geneva lens measure.

The focimeter   The vertex power of a lens may be accurately measured using the focimeter. The axes and major powers of an astigmatic lens may also be measured, as may the power of a prism because most focimeters have a marking device which locates the optical centre of the lens under examination.

The focimeter consists of two main parts—the focusing system and the observation system. The focusing system is composed of an illuminated target which can be moved relative to a standard collimating lens which is part of the instrument (Fig. 7.9). A collimating lens is one which renders light parallel. The target is usually a ring of small dots, formed by a light source placed behind a disc in which are punched a circle of small holes. The lens being tested is placed in a special rack which lies at the principal focus of the collimating lens. The light emerging from the system is viewed through a telescope with an adjustable eye piece. The eyepiece contains a graticule and protractor scale for measuring axis direction of cylindrical lenses. The focimeter measures the vertex power of the lens surface in contact with the lens rest. When examining spectacle lenses, which are designated by their back vertex power, it

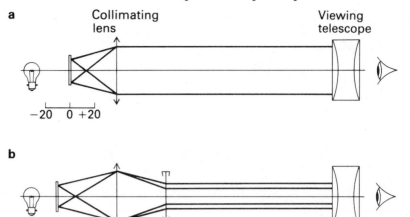

Fig. 7.9 The focimeter.

is thus important to mount the glasses with the back surface of the lens against the rest. Before use, the instrument should be set to zero and the eyepiece adjusted until the dots and the graticule are sharply focused.

Movement of the target allows the vergence of light emerging from the collimating lens to be varied. The target is moved until the light entering the observation telescope is parallel (Fig. 7.9b) in which case a focused image of the target is seen by the observer. The distance through which the target is moved is directly related to the dioptric power of the lens under test. The instrument is so calibrated that the power of the lens can be read off in dioptres.

The image of the target is seen as a ring of dots when a spherical lens is tested. However, in the case of an astigmatic lens the target must be focused separately for the two principal meridians. The dots of the target are then seen as drawn out lines, the length of the lines being proportional to the difference between the two principal powers, that is, the cylindrical power of the lens under test.

To examine an astigmatic lens on the focimeter, e.g. the lens shown in Fig. 7.10a, the instrument is adjusted until one set of line foci is in focus (Fig. 7.10a(i)) and the reading

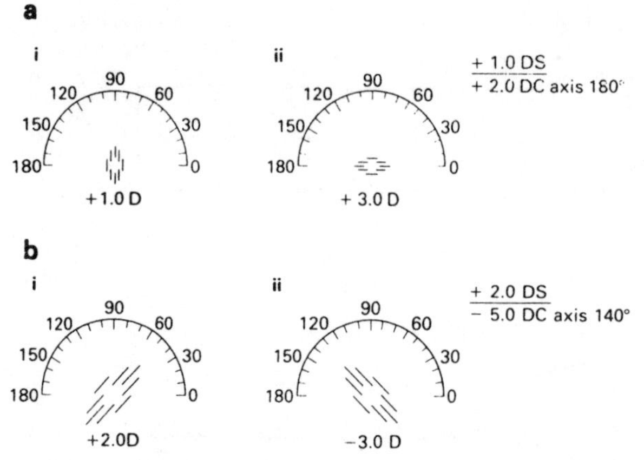

a

i   ii $\dfrac{+1.0\,DS}{+2.0\,DC}$ axis 180°

90   90
120   60   120   60
150   30   150   30
180   0   180   0
+1.0 D   +3.0 D

b

i   ii $\dfrac{+2.0\,DS}{-5.0\,DC}$ axis 140°

90   90
120   60   120   60
150   30   150   30
180   0   180   0
+2.0D   −3.0 D

Fig. 7.10 Use of the focimeter to diagnose astigmatic lenses—image seen.

(+1.0 D) recorded. The instrument is further adjusted until the second set of line foci come into focus (Fig. 7.10a(ii)) and the reading (+3.0 D) and axis (180°) of the lines recorded. The first reading gives the spherical power of the lens. The cylindrical power is calculated by algebraic subtraction of the first reading from the second (+3) − (+1) = +2 D. The axis of the cylinder corresponds to the axis of the second reading, i.e. 180°. Fig. 7.10b shows another example.

A spectacle lens may have a prismatic effect either because it has a prism incorporated in it or because it has been decentred. If there is a prism incorporated in the lens, it will be impossible to bring the image of the target to the centre of the eyepiece graticule. The cross-lines of the graticule are calibrated in intervals of one prism dioptre enabling the prism power to be determined.

In order to detect the prismatic effect of a lens which has been decentred, the centre of the lens should be marked (most machines incorporate a marker). The marked spectacles are then put on to the patient and the degree of decentration of the lens measured by observing the relationship between the centre of the patient's pupil and the optical centre of the lens. Prism power is given by the equation

$$P = F \times D$$

where P is the prismatic power in prism dioptres

F is the lens power in dioptres

D is the decentration in *centimetres*.

Chapter 8

Aberrations of Optical Systems including the Eye*

In practice, the images formed by the various refracting surfaces or systems described in previous chapters fall short of theoretical perfection. Imperfections of image formation are due to several mechanisms, or aberrations, and these have been analysed and means devised to reduce or eliminate their effect. The refracting system of the eye is also subject to aberrations, but there are correcting mechanisms built into the eye itself.

CHROMATIC ABERRATION

When white light is refracted at an optical interface, it is dispersed into its component wavelengths or colours (see Dispersion, Chapter 3, p. 35. Fig. 3.12). The shorter the wavelength of the light, the more it is deviated on refraction. Thus a series of coloured images are formed when white light is incident upon a spherical lens.

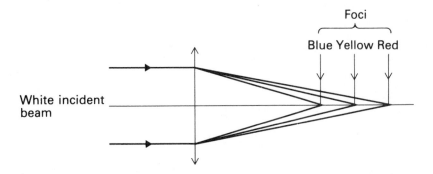

Fig. 8.1 Chromatic aberration.

When lenses are used in instruments, it is desirable to eliminate chromatic aberration.

*The form of the reduced eye (Chapter 9) is used in the diagrams in this chapter.

75

Correction of chromatic aberration

Achromatic lens systems   The dispersive power (Chapter 3, p. 36) of a material is independent of its refractive index. Thus, there are materials of high dispersive power but low refractive index, and vice versa.

Achromatic lens systems are composed of elements (lenses) of varying material combined so that the dispersion is neutralized while the overall refractive power is preserved. (For example, by combining a convex lens of high refractive power and low dispersive power with a concave lens of low refractive power but higher dispersive power, the aberration can be neutralized while preserving most of the convex lens refractive power.) The earliest achromatic lenses were made by combining elements of flint and crown glass.

Ocular chromatic aberration

Refraction by the human eye is also subject to chromatic aberration, total dispersion from the red to the blue image being approximately 1.5–2.0 D. The emmetropic eye focuses for the yellow–green portion of the spectrum. This focused wavelength lies approximately in the middle of the range of retinal sensitivity. Thus approximately 0.75–1.0 D of chromatic aberration lie on either side of the maximally sharp focus (Fig. 8.2).

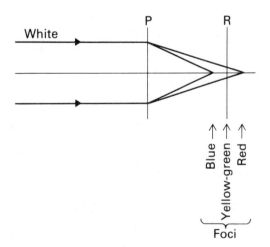

Fig. 8.2 Chromatic aberration, emmetropic eye.

Duochrome test In clinical practice the chromatic aberration of the eye is made use of in the duochrome test. The test consists of two ranks of black Snellen letters, silhouetted against illuminated coloured glass. The upper rank is mounted on red glass, and the lower rank is on green (or blue) glass. The patient views the letters by means of red and green light respectively, and can easily tell which appear clearer. The test is sensitive to an alteration in refraction of 0.25 D or less. A myopic eye sees the red letters more clearly than the green while a hypermetropic eye sees the green letters more distinctly (cf Chapter 10 for definitions of myopia and hypermetropia).

Fig. 8.3   Duochrome test.

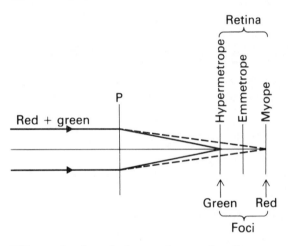

This test is of particular use in ensuring that myopes are not overcorrected.

Colour blindness does not invalidate the test because it depends on the position of the image with respect to the retina. A colour blind patient should be asked whether the upper or lower rank of letters appear clearer.

SPHERICAL ABERRATION

In Chapter 5 the prismatic effect of the peripheral parts of spherical lenses was discussed (Fig. 5.13 and text). It was seen that the prismatic effect of a spherical lens is least in the paraxial zone and increases towards the periphery of the lens. Thus, rays passing through the periphery of the lens are deviated more than those passing through the paraxial zone of the lens.

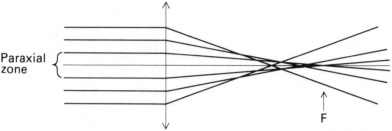

Paraxial
zone

F

Fig. 8.4 Spherical aberration.

Correction of spherical aberration

Spherical aberration may be reduced by occluding the periphery of the lens by the use of 'stops' so that only the paraxial zone is used.

Lens form may also be adjusted to reduce spherical aberration, e.g. plano-convex is better than biconvex. To achieve the best results, spherical surfaces must be abandoned and the lenses ground with *aplanatic surfaces,* that is, the peripheral curvature is less than the central curvature.

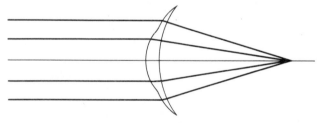

Fig. 8.5   Aplanatic (aspheric) curve to correct spherical aberration.

Another technique of reducing spherical aberration is to employ a doublet. This consists of a principal lens and a somewhat weaker lens of different refractive index cemented together. The weaker lens must be of opposite power, and because it too has spherical aberration, it will reduce the power of the periphery of the principal lens more than the central zone. Usually, such doublets are designed to be both aspheric and achromatic.

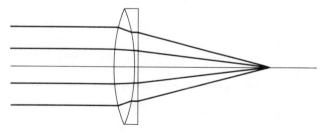

Fig. 8.6   Diagram showing the principle of the aspheric doublet lens.

Ocular spherical aberration

The effect of spherical aberration in the human eye is reduced by several factors.

1. The anterior corneal surface is flatter peripherally than at its centre, and therefore acts as an aplanatic surface.
2. The nucleus of the lens of the eye has a higher refractive index than the lens cortex (Chapter 9). Thus the axial zone of the lens has greater refractive power than the periphery.

Fig. 8.7 Ocular spherical aberration (SA)—compensatory mechanisms.

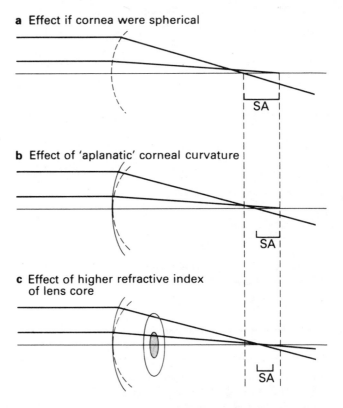

a Effect if cornea were spherical

b Effect of 'aplanatic' corneal curvature

c Effect of higher refractive index of lens core

3. Furthermore, in the eye, the iris acts as a stop to reduce spherical aberration. The impairment of visual acuity that occurs when the pupil is dilated is almost entirely due to spherical aberration. Optimum pupil size is 2–2.5 mm.
4. Finally, the retinal cones are much more sensitive to light which enters the eye paraxially than to light which enters obliquely through the peripheral cornea (Stiles–

Crawford effect). This directional sensitivity of the cone photoreceptors limits the visual effects of the residual spherical aberration in the eye.

OBLIQUE ASTIGMATISM

Oblique astigmatism is an aberration which occurs when rays of light traverse a spherical lens obliquely. When a pencil of light strikes the lens surfaces obliquely a toric effect is introduced. The emerging rays form a Sturm's conoid with two line foci.

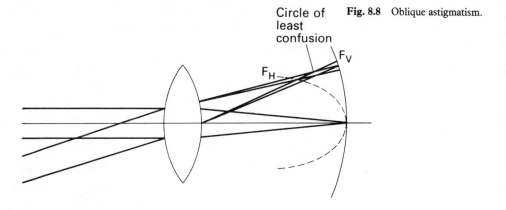

Circle of least confusion

F_H F_V

Fig. 8.8   Oblique astigmatism.

F_H and F_V represent the horizontal and vertical line foci of a Sturm's conoid.

Correction of oblique astigmatism

To eliminate oblique astigmatism the lens should be orientated so that incident light is parallel to the principal axis. If this is not possible the aberration may be reduced by restricting the aperture of the lens.

Furthermore, oblique astigmatism is considerably affected by the form of the lens used. It is much worse in biconvex and biconcave lenses than in meniscus lenses.

Calculations have been made and tables compiled indicating the optimum form of single lenses for reducing both spherical and oblique aberrations. Such lenses are known as *best form lenses*, and they are usually in meniscus form.

Ocular oblique astigmatism

This aberration occurs in the human eye but its visual effect is minimal. The factors which reduce ocular oblique astigmatism are as follows:

1. The aplanatic curvature of the cornea reduces oblique astigmatism as well as spherical aberration.
2. The retina is not a plane surface, but a spherical surface. In practice the radius of curvature of the retina in the emmetropic eye means that the circle of least confusion of the Sturm's conoid formed by oblique astigmatism (Fig. 8.8) falls on the retina.
3. Finally, the astigmatic image falls on peripheral retina which has relatively poor resolving power compared to the retina at the macula. Visual appreciation of the astigmatic image is therefore limited.

COMA

Coma is really spherical aberration applied to light coming from points not lying on the principal axis. Rays passing through the periphery of the lens are deviated more than the central rays and come to a focus nearer the principal axis (Fig. 8.9). This results in unequal magnification of the image formed by different zones of the lens. The composite image is not circular but elongated like a coma or comet.

Fig. 8.9 Coma aberration.

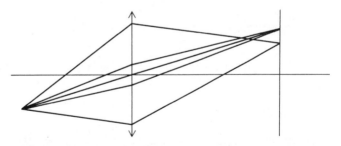

Correction of coma aberration

As in the case of oblique astigmatism this aberration can be avoided by limiting rays to the axial area of the lens, and by using the principal axis of the lens rather than a subsidiary axis.

Ocular coma aberration is not of practical importance for the reasons given under oblique astigmatism.

IMAGE DISTORTION

When an extended object is viewed through a spherical lens, the edges of the object, viewed through the peripheral zones of the lens are distorted. This is due to the increased prismatic effect of the periphery of the lens which produces uneven magnification of the object. A concave lens causes 'barrel' distortion while a convex lens causes 'pin-cushion' distortion. These effects prove a real nuisance to wearers of high-power spectacle lenses, e.g. aphakic patients.

Fig. 8.10   Image distortion.

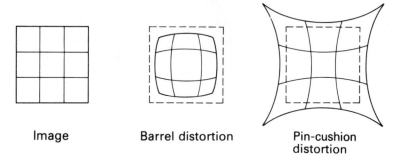

Image   Barrel distortion   Pin-cushion
                                                  distortion

CURVATURE OF FIELD

The term curvature of field indicates that a plane object gives rise to a curved image (Fig. 8.11).

Fig. 8.11   Curvature of field.

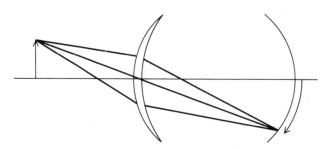

This occurs even when spherical aberration, oblique astigmatism and coma have been eliminated. The effect is dependant upon the refractive index of the lens material and the curvature of the lens surfaces.

Ocular curvature of field

In the eye the curvature of the retina compensates for curvature of field.

Chapter 9   **Refraction by the Eye**

The thin lens formula (Chapter 5) is inadequate to deal with the refracting system of the eye, which is composed of a number of refracting surfaces separated by relatively long distances.

However, the theory and formula of refraction by thick lenses can be directly applied to the eye. Refraction by thick lenses is therefore considered below, and the principles subsequently applied to the eye.

THICK LENS THEORY

Cardinal points

It will be recalled that the thin lens formula ignores lens thickness and considers refraction only at the two lens surfaces (Chapter 5). For a thick lens this approach is invalidated by the greater separation of the two refracting surfaces by the lens substance. The full mathematical analysis of refraction by a thick lens is very complex. It has been simplified by the introduction of the concept of *principal points* and *principal planes*. These are hypothetical planes and points such that a ray incident at the first principal point or plane, P_1, leaves the second principal point or plane, P_2, at the same vertical distance from the principal axis (Fig. 9.1). The exact position of the principal point is calculated from the curvatures of the lens surfaces, the lens thickness and the refractive index of the

Fig. 9.1 Thick lens. Cardinal points.

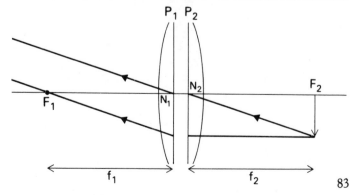

lens material. The principal planes intersect the principal axis at right angles at the principal points.

There are two further points, the *nodal points*, N_1 and N_2, which correspond to the centre of a thin lens. Any ray directed towards the first nodal point, N_1, leaves the lens as if from the second nodal point, N_2, and parallel with its original direction, i.e. undeviated. When the medium on both sides of the thick lens is the same, the nodal points coincide with the principal points. When the media on opposite sides of the lens are different, the nodal points do not coincide with the principal points.

The *principal foci*, F_1 and F_2, have the same meaning as for a thin lens (Chapter 5), the focal lengths, f_1 and f_2, being measured from the principal points (see below).

Vertex and equivalent power of thick lenses

a Concave meniscus

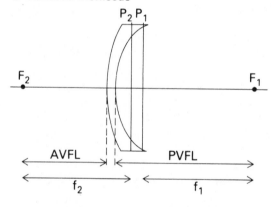

b Convex meniscus

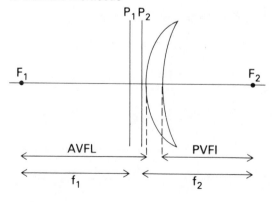

Fig. 9.2. Thick lens—true and vertex focal lengths.

Figs. 9.1 and 9.2 show that the principal points do not lie on the surface of the lens. They may lie within the lens, or in meniscus form lenses, outside the lens substance. In practice, the measurement given by instruments such as the focimeter is the distance of the principal focus from the central surface or vertex of the lens. This distance is the anterior or posterior *vertex focal length*, AVFL or PVFL, and it must be distinguished from the focal length, f_1 or f_2. Furthermore note that the anterior and posterior vertex focal lengths are not equal to each other (Fig. 9.2).

The reciprocal of the posterior vertex focal length, expressed in metres, is the posterior *vertex power*, expressed in dioptres. An alternative term is the *back vertex power*. This value differs from the 'true' focal power or *equivalent power* of the lens. The *equivalent power* of a thick lens is calculated from the two surface powers plus a correction for vergence change due to lens thickness. The discrepancy between equivalent and back vertex power may be a cause of error in dispensing high-powered spectacle lenses, or highly curved contact lenses. However, lens manufacturers and dispensing opticians are aware of this problem and mathematical tables exist which allow the appropriate adjustment to be made.

Spectacle lenses are graded by their back vertex power as it is their posterior vertex focal length which is relevant to the correction of optical defects in the eye. The second principal focus of the lens must correspond to the far point of the eye if a clear retinal image is to be formed. This is fully explained and illustrated in Chapter 10. Back vertex power should not be confused with back vertex distance, which is merely the distance between the eye and the back vertex of a spectacle lens.

REFRACTION BY THE EYE

The foregoing analysis of refraction by a thick lens is just one application of the Gaussian theory of cardinal points. The theory may be applied to any system of coaxial spherical refracting surfaces, including the human eye.

There are three major refracting interfaces to be considered in the eye—the anterior corneal surface and the two surfaces of the lens. The effect of the posterior corneal surface is very small compared with these three as the difference in refractive index between corneal stroma and aqueous is not large (Table 9.1).

(Air	1.000)
Cornea	1.376
Aqueous humour	1.336
Lens (cortex-core)	1.386–1.406
Vitreous humour	1.336

Table 9.1   Refractive indices of the transparent media of the eye (Gullstrand).

In order to calculate the cardinal points, the radii of curvature, and distances separating the refracting surfaces must also be known. These have been determined experimentally by several observers. As in the case of any anatomical measurement there is some physiological variation and the values given are the means. The results of Gullstrand are given here (Tables 9.1, 9.2, 9.3) as it is on these that the given calculations of the schematic and reduced eye are based. These measurements are known as the *optical constants* of the eye. However, there are several other sets of measurements available from other observers which differ slightly one from another. No one set is in general standard use.

Cornea, anterior surface	0
Cornea, posterior surface	0.5
Lens, anterior surface	3.6
Lens, posterior surface	7.2
Lens core, anterior surface	4.146*
Lens core, posterior surface	6.565*

*Calculated values

Table 9.2   Position of refracting surfaces of the eye (in mm behind anterior corneal surface) (Gullstrand).

Cornea, anterior surface	7.7
Cornea, posterior surface	6.8
Lens, anterior surface	10.0
Lens, posterior surface	−6.0
Lens core, anterior surface	7.911*
Lens core, posterior surface	−5.76*

*Calculated values

Table 9.3   Radii of curvature of refracting surfaces of the eye (in mm) (Gullstrand).

The schematic eye

In the schematic eye, as described by Gullstrand, the refracting system is expressed in terms of its cardinal points (measured in mm behind the anterior corneal surface).

Note that the nodal points, via which rays of light pass

Fig. 9.3 The schematic eye.

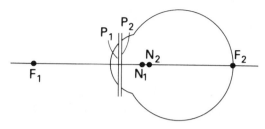

Table 9.4 Schematic eye, cardinal points (distance in mm behind anterior corneal surface) (Gullstrand).

First principal point P_1	1.35
Second principal point P_2	1.60
First nodal point N_1	7.08*
Second nodal point N_2	7.33*
First focal point	−15.7
Second focal point	24.4
Refractive power	+58.64 D

*Calculated by Percival (1928)

undeviated (page 84), are removed from the principal points, which lie at the intersection of the principal planes with the principal axis. This is because the refracting media on each side of the refracting system of the eye are different, namely air ($n = 1$) and vitreous ($n = 1.336$).

The nodal points straddle the posterior pole of the crystalline lens. The pupil of the eye allows only a relatively small paraxial pencil of light to enter the eye. Such paraxial rays are refracted and concentrated through the nodal points and adjacent posterior lens substance. Therefore even a small posterior polar cataract produces gross impairment of vision when the pupil is small.

The reduced eye

Matters were simplified further by Listing (1853) who chose a single principal point lying midway between the two principal points of the schematic eye. A single nodal point was postulated in the same way, and the focal lengths adjusted with reference to the new principal point. The result is the *reduced eye,* in which the eye is treated as a single refracting surface of power +58.6 D.

Fig. 9.4 The reduced eye.

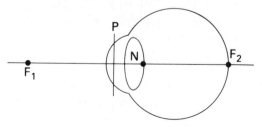

Principal Point P	1.35
Nodal point N	7.08
First focal point	−15.7
Second focal point	24.13

Table 9.5   The reduced eye (distances in mm behind anterior corneal surface) (after Gullstrand).

Many sets of figures exist for the reduced eye. Those quoted above are based on Gullstrand's data (Duke-Elder, *System of Ophthalmology*, Vol V, p. 121).

It must be emphasized that the distances given for the focal points are measured from the anterior corneal surface. The corresponding focal lengths, measured from the principal point, are −17.05 mm and 22.78 mm.

Note that the nodal point lies in the posterior part of the lens, while the second focal point lies 24.13 mm behind the cornea, i.e. on the retina of the normal eye.

The anterior focal length of the aphakic eye was found experimentally to be −23.23 mm (the first focal point lying 21.88 mm in front of the cornea). This gives a calculated power of +43 D for the aphakic eye. The crystalline lens thus has an effective power *in situ* of +15 D (the difference between the power of the phakic and aphakic eye [58 D–43 D]). Its actual power, taken in isolation from the other refractive elements of the eye is given by Gullstrand as +19 D. The discrepancy results from the fact that *in vivo* the lens is only one element in a larger refracting system.

The cornea is the only refracting element remaining in the aphakic eye. From the above figures it thus has a power of 43 D, three times as powerful as the crystalline lens in the intact eye. The relatively greater power of the cornea is due to the greater difference in refractive index between air (1.000) and cornea (1.376), as compared to aqueous and vitreous humour (1.336) and lens (1.406), i.e. a difference of 0.376 as opposed to 0.140 (0.070 × 2 because both the anterior and the posterior lens surfaces contribute to the power of the crystalline lens).

A dramatic example of the importance of the air/cornea interface occurs when a swimmer opens his eyes under water. He finds his vision is blurred. The difference in refractive index between water and cornea is only 0.040 (i.e. cornea 1.376—water 1.336). This problem is eliminated by the swimmer keeping air in front of the cornea by wearing goggles.

The form of the reduced eye is used in all the subsequent discussion of the optics of the eye. The student is advised to commit the values to memory.

Reduced eye—construction of retinal image

Using the reduced eye, it is simple to construct the retinal image formed under various conditions (see Chapter 10).

The reduced eye itself is represented by two parallel lines, which indicate the principal plane, P, and the retina, R. These intersect the principal axis (optical axis) at right angles. The nodal point, N, is indicated by a point, as is the anterior focus, F_a. The second principal focus, F_2 falls on the retina in the emmetropic eye (Fig. 9.5).

Two rays are used to construct the image formed by parallel light incident upon the eye:

1. A ray passing through the anterior focus, F_a, which after refraction at the principal plane, P, continues parallel to the principal axis.
2. A ray passing through the nodal point, N, undeviated.

Fig. 9.5 Reduced eye—image formation.

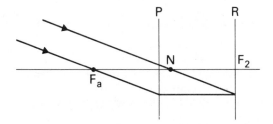

The size of the retinal image may also be calculated from the same construction.

Fig. 9.6 Reduced eye—retinal image size.

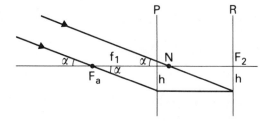

It can be seen from Fig. 9.6 that the light subtends angle α at the nodal point, N, as well as at the anterior focus, F_a. *Retinal image size is therefore directly related to the angle subtended by an object at the nodal point* (f_1 being constant in the individual eye).

Because $\tan \alpha = h/f_1$
the retinal image size $h = \tan \alpha \cdot f_1$

The angle, α, subtended by an object at the nodal point is called the *visual angle* (Figs. 9.6 and 9.7).

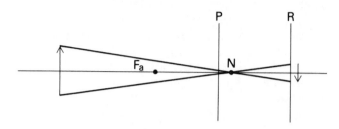

Fig. 9.7 Visual angle—near object.

It will be appreciated that as an object of given size approaches the eye, it subtends a greater visual angle and thus appears larger (cf. the Loupe, Chapter 5).

Variable states of emmetropia

In the emmetropic eye, the second principal focus falls on the retina. Parallel incident light is therefore focused on the retina, without accommodative effort.

The emmetropic state is compatible with a range of refractive powers, if the axial length of the eye is appropriate to its dioptric power (Fig. 9.8).

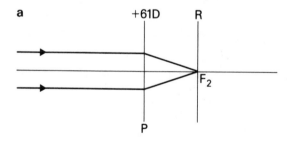

Fig. 9.8 Variable states of emmetropia. (See Fig. 9.8c on facing page.)

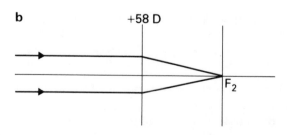

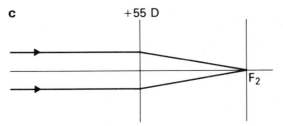

c +55 D

F₂

Accommodation of the eye

If the refractive power of an emmetropic eye were fixed and unalterable, only objects at infinity would be clearly seen. Light from nearer objects would be brought to a focus beyond the second principal focus, F_2, (see convex lenses, Chapter 5) and no clear image would be formed on the retina.

This problem is overcome by the ability of the eye to increase its dioptric power. The crystalline lens is held suspended under tension by the suspensory ligament which attaches it to the ring of ciliary muscle. Ciliary muscle contraction reduces the tension on the suspensory ligament and lens, allowing the lens to assume a more globular shape (Fig. 9.9). The curvatures of the lens surfaces and the lens thickness are increased and thus the dioptric power is increased. Most of the change in curvature occurs at the anterior lens surface, which moves forwards slightly towards the cornea. This ability of the eye to increase its dioptric power is called *accommodation*.

Fig. 9.9 Accommodation— diagram illustrating change in crystalline lens form.

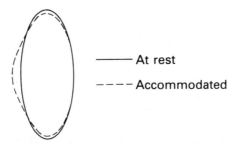

———— At rest

– – – – Accommodated

Before analysing the process of accommodation further it is necessary to define certain terms:
1. The *far point* of distinct vision is the position of an object such that its image falls on the retina in the relaxed eye, i.e. in the absence of accommodation. The far point of the emmetropic eye is at infinity.

2. The *near point* of distinct vision is the nearest point at which an object can be clearly seen when maximum accommodation is used.
3. The *range of accommodation* is the distance between the far point and the near point.
4. The *amplitude of accommodation* is the difference in dioptric power between the eye at rest and the fully accommodated eye.
5. The dioptric power of the resting eye is called its *static* refraction.
6. The dioptric power of the accommodated eye is called its *dynamic* refraction.

Mathematically, the amplitude of accommodation can be calculated from the reciprocals of the near and far point distances measured in metres. These are the *dioptric values* of the near and far point distances. The amplitude of accommodation is given by the formula

$$A = P - R$$

where A is the amplitude of accommodation in dioptres
     P is the dioptric value of the near point distance
     R is the dioptric value of the far point distance.

Applying this formula to the case of an emmetropic eye with a near point of 10 cm.

$$P = 10 \text{ D (the reciprocal of 0.1 m)}$$
$$R = 0 \text{ (the reciprocal of infinity is zero)}$$
therefore $A = 10$ D.

To calculate the accommodative power required to focus an intermediate point within the range of accommodation the formula is amended to

$$A = V - R$$

where A is the accommodative power required, in dioptres
     V is the dioptric value of the intermediate point
     R is the dioptric value of the far point (the far point distance in hypermetropia, being behind the eye, carries a negative sign).

Thus, to focus an object at 1 m, the emmetropic eye must exert one dioptre of accommodative power ($A = 1 - 0$).

The amplitude of accommodation declines with advancing age, giving rise to the condition of presbyopia—the inability to focus near objects. This requires spectacle correction and will be discussed later (Chapter 10).

Information regarding the changes in lens form during accommodation has come from studies of the catoptric images of the eye.

Catoptric images

The changes occurring in the lens during accommodation are inaccessible to direct measurement, and therefore an indirect method must be employed. Similarly the optical constants of the eye (Tables 9.1–9.3) must also be measured indirectly.

Each refracting interface in the eye also acts as a spherical mirror, reflecting a small portion of the light incident upon it (Chapter 2). Four images are therefore formed by reflection at the four interfaces—the anterior and posterior corneal surfaces and the anterior and posterior lens surfaces.

These images are called *catoptric images* or *Purkinje–Sanson images*. The latter name honours Purkinje who first described them and Sanson who first used them for diagnostic purposes.

Fig. 9.10 Purkinje–Sanson images. Apparent positions as seen by the observer.

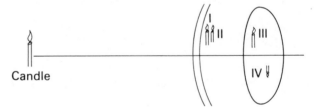

Images I, II and III (from anterior corneal, posterior corneal and anterior lens surfaces respectively) are erect, virtual images because they are formed by convex reflecting surfaces (see Chapter 2).

Image IV (from the posterior lens surface) is a real, inverted image because it is formed by a concave reflecting surface (see Chapter 2).

When using images II, III and IV to make measurements, account must be taken of refraction of the reflected light as it re-emerges from the eye (cf. real and apparent depth, Chapter 3). Fig. 9.10 shows the apparent positions of the images when viewed by the observer. However, the actual images lie deeper in the eye—image I lies just behind the anterior lens capsule, and image II is close behind it. Image III is located in the vitreous, and image IV, as it comes from the concave posterior lens surface, is inverted and in the anterior lens substance.

Use of the first image (image I) to study the anterior corneal curvature is routine in clinical practice. The regularity of the curvature is examined using Placido's disc, and the radius of curvature is measured using the keratometer. These instruments are described in Chapter 13.

The first image is also used in the diagnosis and measurement of squint.

Much information regarding the changes in lens form during accommodation has also been obtained from the study of images III and IV.

Chapter 10   **Optics of Ametropia**

In contrast to emmetropia (Chapter 9, p. 90) the ametropic eye fails to bring parallel light to a focus on the retina, i.e. the second principal focus of the eye does not fall on the retina.

Myopia

In the myopic eye, the second principal focus lies in front of the retina.

Fig. 10.1   Myopia.

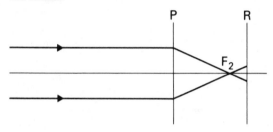

This may be because the eye is abnormally long. This is called *axial myopia* and includes high myopia in which there may be a posterior staphyloma.

Alternatively, the eye may be of normal length, but the dioptric power may be increased. This is called *refractive or index myopia*. Examples of this are keratoconus, where the corneal refractive power is increased, and nucleosclerosis, where the refractive power of the lens increases as the nucleus becomes more dense.

Hypermetropia

In the hypermetropic eye, the second principal focus lies behind the retina.

Fig. 10.2   Hypermetropia.

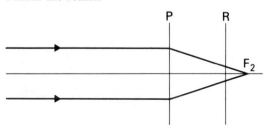

If the eye is short relative to its focal power, then *axial hypermetropia* results.

Alternatively, if the refractive power of the eye is inadequate then *refractive hypermetropia* results. Aphakia is an extreme example of refractive hypermetropia.

Phakic patients can overcome some or all of their hypermetropia by using accommodation for distance vision. They then have to exercise extra accommodation for near vision. Because the amplitude of accommodation declines with age (Chapter 11), these patients require reading glasses at a younger age than emmetropic patients.

Hypermetropia is classified into manifest and latent hypermetropia. *Manifest* hypermetropia is defined as the strongest convex lens correction accepted for clear distance vision. *Latent* hypermetropia is the remainder of the hypermetropia which is masked by ciliary tone and involuntary accommodation. This may account for several dioptres, especially in children, for whom mydriatic refraction is necessary to ascertain the full magnitude of the refractive error.

Hypermetropia which can be overcome by accommodation is called *facultative*, while hypermetropia in excess of the amplitude of accommodation is called *absolute*.

Astigmatism

The refractive power of the astigmatic eye varies in different meridians. The image is formed as a Sturm's conoid (cf. Fig. 6.9).

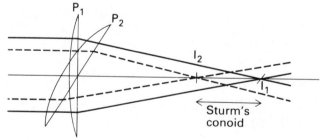

Fig. 10.3 Astigmatism—image formation.

Fig. 10.4 Astigmatism—classification. Image position relative to retina.

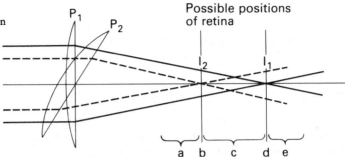

Retina a = Compound hypermetropic astigmatism — rays in all meridians come to a focus behind the retina.

Retina b = Simple hypermetropic astigmatism — rays in one meridian focus on the retina, the other focus lies behind the retina.

Retina c = Mixed astigmatism — one line focus lies in front of the retina, the other behind the retina.

Retina d = Simple myopic astigmatism — one line focus lies on the retina, the other focus lies in front of the retina.

Retina e = Compound myopic astigmatism — rays in all meridians come to a focus in front of the retina.

Anisometropia

When the refraction of the two eyes is different, the condition is known as *anisometropia*. Small degrees of anisometropia are commonplace. Larger degrees are a significant cause of amblyopia. A disparity of more than 1 D in the hypermetropic patient is enough to cause amblyopia of the more hypermetropic eye because accommodation is a binocular function, i.e. the individual eyes cannot accommodate by different amounts. The more hypermetropic eye therefore remains out of focus. The myopic patient with anisometropia is less likely to develop amblyopia because both eyes have clear near vision. However, when one eye is highly myopic it usually becomes amblyopic.

Older patients with nucleosclerosis and resulting index myopia affecting one eye more than the other may not tolerate the full spectacle correction of the more myopic eye as they are not accustomed to coping with aniso-

metropia. However, myopic patients who have been aniso-
metropic all their lives may tolerate higher degrees of
anisometropia and achieve binocular vision with more
than 2 D difference between the two eyes.

Pin-hole test

Because no focused image is formed on the retina of the
ametropic eye, the visual acuity is reduced. The pin-hole
test is a useful method of determining whether reduced
visual acuity is due to refractive error rather than ocular
pathology or neurological disease.

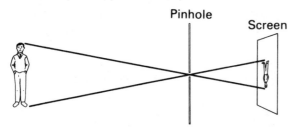

Fig. 10.5 Optical principle of
the pin hole.

The pin-hole theoretically allows only one ray from each
point on an object to pass through to the screen. Thus, a
clear image is formed regardless of the position of the
screen. Likewise the use of the ideal pin-hole leads to the
formation of a clear retinal image irrespective of the
refractive state of the eye.

But in practice the pin-holes available clinically allow a
narrow pencil of light to pass through them, rather than a
single ray.

Fig. 10.6 (on facing page) shows that in low degrees of
refractive error the pin-hole's effect is sufficient to improve
the clarity of the retinal image to such an extent that a good
visual acuity results. However, in high degrees of ametropia, although the pin-hole helps, the retinal image is still
too diffuse to achieve the improvement that is found in the
case of low refractive errors. Thus errors outside the range
+4 D to –4 D sphere are not corrected to 6/6 with a pin-
hole.

Stenopaeic slit

The stenopaeic slit can be used to determine the refraction and principal axes in astigmatism. The slit aperture
acts as an elongated 'pin-hole', only allowing light in the

Fig. 10.6 Pin hole—practical effect in ametropia.

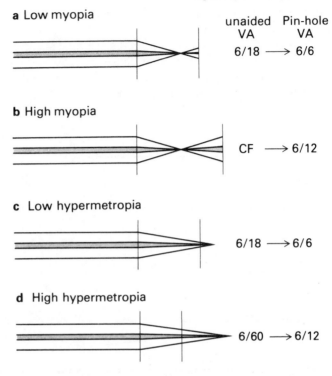

axis of the slit to enter the eye. Hence, when the slit lies in one principal axis of the astigmatic eye, the second line focus is eliminated and the blur of Sturm's conoid reduced thus allowing a clearer image to be formed.

During the refraction of a patient with astigmatism, the slit is first rotated to a position in which the clearest vision is obtained. Spherical lenses are added to give further improvement in acuity. The slit is then rotated through 90° and the spherical lens power adjusted to give best subjective acuity. The cylindrical correction required by the eye equals the algebraic difference between the two spherical corrections used, and its axis is that of the original direction of the slit.

In cases of corneal scarring, the stenopaeic slit may be used to determine the meridian along which the cornea is least deformed. In those cases where an optical iridectomy is indicated, this should be performed in this meridian.

Far point

The far point (FP) of an eye is the position of an object such that its image falls on the retina of the relaxed eye, i.e. in the absence of accommodation.

The distance of the far point from the principal plane of the eye is denoted by r, which according to sign convention carries a negative sign in front of the principal plane and a positive sign behind the principal plane.

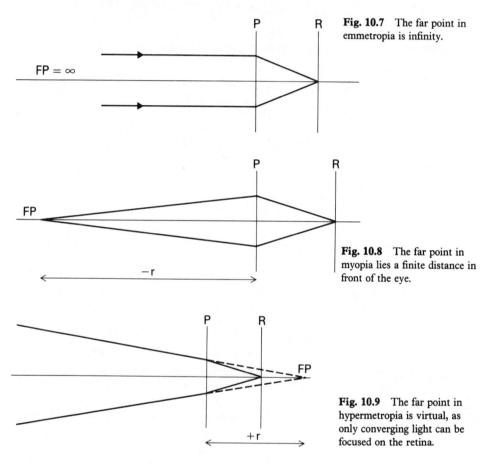

Fig. 10.7   The far point in emmetropia is infinity.

Fig. 10.8   The far point in myopia lies a finite distance in front of the eye.

Fig. 10.9   The far point in hypermetropia is virtual, as only converging light can be focused on the retina.

Optical correction of ametropia

The purpose of the correcting lens in ametropia is to deviate parallel incident light so that it appears to come from the far point in myopia or to be converging towards the virtual far point in hypermetropia. The light will then be brought to a focus by the eye on the retina. *Thus the far point of the eye must coincide with the focal point of the lens.*

Fig. 10.10 Correction of hypermetropia.

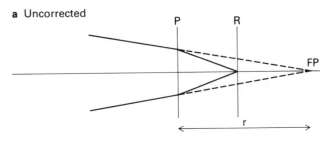

a Uncorrected

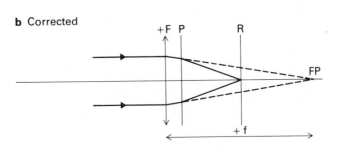

b Corrected

The focal length, f, of the correcting lens is approximately equal to the distance, r, of the far point from the principal plane when the correcting lens is close to the principal plane (Fig. 10.10). Thus the power of lens, F, required is

$$F = \frac{1}{f} \approx \frac{1}{r}$$

where F is the power of the lens in dioptres

f is the focal length of the lens in metres

r is the distance of the far point from the principal plane in metres.

Again $-r \approx -f$

and $-F = \frac{1}{-f} \approx \frac{1}{-r}$

The reciprocal of the far point distance r, in metres, is symbolized by R, expressed in dioptres. R is known as the *static refraction* or the *ametropic error*.

In practice, the correcting lens in ametropia is usually held in spectacles. The lens is, therefore, some distance in front of the principal plane of the eye. The power of the

a Uncorrected

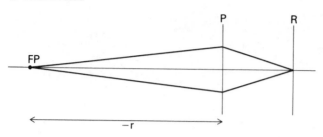

Fig. 10.11   Correction of myopia.

b Corrected

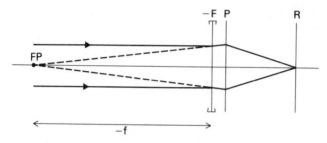

lens necessary to correct a specific degree of ametropia must therefore be adjusted so that the far point and the focus of the lens still coincide (see below—effective power of lenses).

Effective power of lenses

If a correcting lens is moved either towards or away from the eye, its vergence power at the principal plane of the eye changes. The focus of the lens and the far point of the eye no longer coincide (Figs. 10.12 and 10.13, opposite).

It can also be seen from the diagrams that on moving either a convex or a concave lens away from the eye, the image is moved forward.

In the uncorrected hypermetropic eye (Fig. 10.2) the image falls behind the retina. The purpose of the correcting convex lens is to bring the image forward on to the retina. When the correcting lens is moved further away from the eye the image is brought still further forward. Thus the effectivity of the lens is said to be increased. Therefore in this position a weaker convex lens throws the

Fig. 10.12 Diagram showing the change in effectivity of a convex lens on moving it away from the eye.

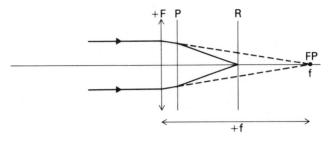

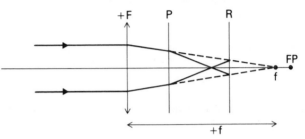

Fig. 10.13 Diagram showing the change in effectivity of a concave lens on moving it away from the eye.

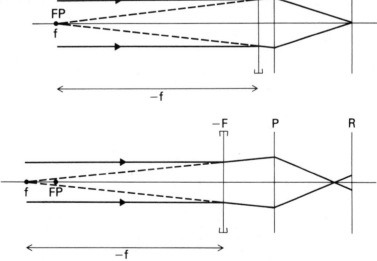

image onto the retina and corrects the hypermetropia.

In the myopic eye (Fig. 10.1) the image falls in front of the retina. The purpose of the correcting concave lens is to take the image back on to the retina. When the correcting lens is moved further away from the eye, the image moves forward again. Thus the effectivity of the lens is said to be reduced. Therefore, in this position, a stronger concave lens is needed to throw the image onto the retina.

In practice, patients with strong convex lenses, especially those who are aphakic, sometimes pull their glasses down their nose in order to read. The enhanced effectivity thus produced is sufficient to provide the reading correction. Also, myopes dislike their glasses slipping down their nose (as they tend to do if heavy glass lenses are used) as this makes the correction less effective.

Thus, to correct a specific degree of ametropia, the power of the correcting lens must be adjusted to take into account its position in front of the eye. A general formula exists for this purpose and applies to both convex and concave lenses.

Suppose a lens of focal length f_1 at a given position in front of the ametropic eye, corrects the refractive error. Then a different lens of focal length $(f_1 - d)$ is required when the correction is moved a distance d towards or away from the eye (Figs. 10.14 and 10.15). The value of d is positive if the lens is moved towards the eye, and negative if moved away from the eye. The usual sign convention applies to the lens.

Thus $F_2 = \dfrac{1}{f_1 - d}$

where F_2 is the power of lens in dioptres required at the new position
   f_1 is the focal length in metres of the original lens
and   d is the distance moved in metres.

Mathematically the above formula can also be expressed as

$$F_2 = \dfrac{F_1}{1 - dF_1}$$

where F_1 is the dioptric power of the original lens.

Practical application—Back Vertex Distance

For any lens of power greater than 5 dioptres, the position in front of the eye materially affects the optical correction of ametropia. This is especially true in aphakia where high power lenses are prescribed. For this reason the refractionist must state how far in front of the eye the trial lens is situated so that the dispensing optician can adjust the lens power if a contact lens is to be used, or if spectacles are to be worn at a different distance, e.g. because of a high-bridged nose or deep-set eyes.

Therefore any high powered lens should be placed in the back cell of the trial frame and the distance between the back of the lens and the cornea measured. This is called the *Back Vertex Distance* (BVD) and must be given with all prescriptions over 5 dioptres. The measurement may be made with a ruler held parallel to the arm of the trial frame. Other means include a small rule which is slipped through a stenopaeic slit placed in the back cell of the trial frame until it touches the closed eyelid. Two millimetres must be added to the measurement to correct for the thickness of the lid.

Fig. 10.14 Correction of hypermetropia.

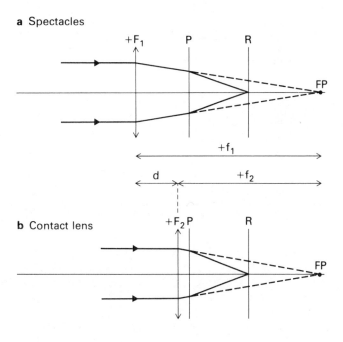

a Spectacles

b Contact lens

Example 1
Refraction shows that an aphakic patient requires a
+10.0 D lens at BVD 15 mm. He needs a contact lens (F_2)

$$F_2 = \frac{F_1}{1-dF_1}$$

Required power of

$$\begin{aligned}
\text{contact lens} = F_2 &= \frac{+10}{1 - 0.015 \times 10} \\
&= \frac{+10}{1 - 0.15} \\
&= \frac{+10}{0.85} \\
&= +11.75 \text{ D}
\end{aligned}$$

a Spectacles

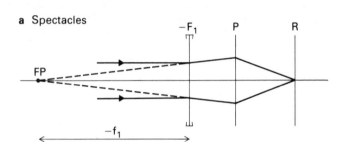

b Contact lens

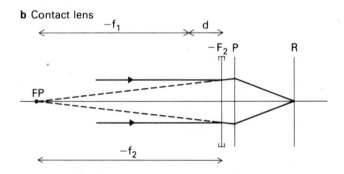

Fig. 10.15 Correction of myopia.

Example 2
Likewise a high myope, whose spectacle correction is
−10.0 D at BVD 14 mm requires a contact lens (F_2)

$$F_2 = \frac{F_1}{1-dF_1}$$

$$\text{Power of contact lens} = F_2 = \frac{-10}{1 - (+0.014 \times [-10])}$$

$$= \frac{-10}{1 - (-0.14)}$$

$$= \frac{-10}{1 + 0.14}$$

$$= \frac{-10}{1.14}$$

$$= -8.75 \text{ D}$$

Happily, tables exist which give the value of F_2 when F_1 and d are known.

Spectacle magnification

The optical correction of ametropia is associated with a change in the retinal image size. The ratio between the corrected and uncorrected image size is known as the *Spectacle magnification*.

$$\text{Spectacle magnification} = \frac{\text{corrected image size}}{\text{uncorrected image size}}$$

Clinically, it is more useful to compare the corrected ametropic image size with the emmetropic image size. This ratio is known as the *Relative Spectacle Magnification* (RSM).

$$\text{Relative Spectacle Magnification} = \frac{\text{corrected ametropic image size}}{\text{emmetropic image size}}$$

In axial ametropia, if the correcting lens is placed at the anterior focal point of the eye, the image size is the same as in emmetropia. The RSM is therefore unity.

Axial hypermetropia RSM = 1

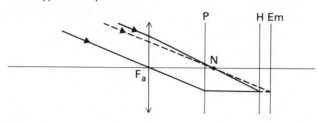

Fig. 10.16   Relative spectacle magnification—axial ametropia with correcting lens at anterior focal point of the eye.

Axial myopia RSM = 1

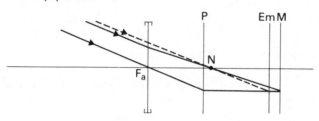

However, in axial myopia, if the correcting lens is worn nearer to the eye than the anterior focal point, the image size is increased. The relative spectacle magnification is therefore greater than unity. Contact lenses in axial myopia thus have a magnifying effect.

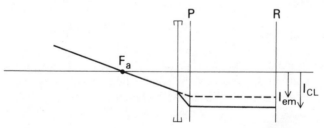

Fig. 10.17   Relative spectacle magnification—axial myopia with correcting lens nearer the eye than the anterior focal point (contact lens).

where I_{em} is the image size when correction is at the anterior focal point, which equals the emmetropic image size (Fig. 10.16)

I_{CL} is the image size when correction is closer to the eye than the anterior focal point.

In contrast to axial ametropia, the image size in refractive ametropia differs from the emmetropic image size even when the correcting lens is at the anterior focal point of the eye. The image size in refractive hypermetropia is increased, thus the relative spectacle magnification is greater than unity. In refractive myopia the image size is diminished, and thus the relative spectacle magnification is less than unity (Fig. 10.18).

Fig. 10.18 Relative spectacle magnification—refractive ametropia with correcting lens at anterior focal point of the eye.

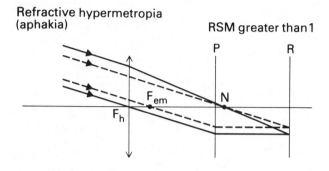

Refractive hypermetropia (aphakia)

RSM greater than 1

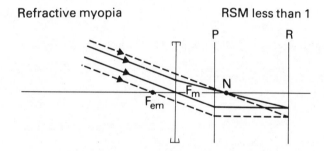

Refractive myopia

RSM less than 1

Furthermore, in refractive ametropia, if the correcting lens is worn nearer to the eye than the anterior focal point, the image size approaches the emmetropic image size. The relative spectacle magnification thus approaches unity.

Spectacle correction in aphakia (refractive hypermetropia) produces a relative spectacle magnification of 1.36 when placed at the anterior focal point of the aphakic eye (23.2 mm in front of the principal plane) (see p. 80). However, when a contact lens is used, the relative spectacle magnification is reduced to 1.1.

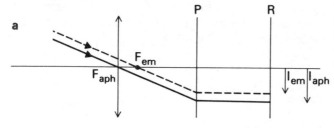

Fig. 10.19 Relative spectacle magnification—correction of aphakia with (a) spectacles at the anterior focal point, (b) contact lens.

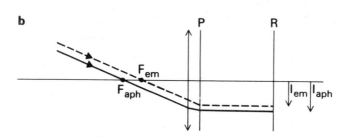

Spectacles are usually worn 12–15 mm in front of the cornea and the aphakic relative spectacle magnification at this position is approximately 1.33.

Calculation of RSM in aphakia when correcting lens is at the anterior focal point of the aphakic eye.

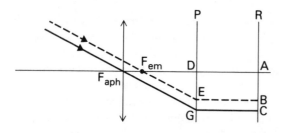

Fig. 10.20 Correction of aphakia with spectacles at anterior focal point.

Emmetropic anterior focal length, $F_{em}D = 17.05$ mm
Aphakic anterior focal length $F_{aph}D = 23.23$ mm

AB = DE = emmetropic image size
AC = DG = corrected aphakic image size

$$RSM = \frac{AC}{AB} = \frac{DG}{DE}$$

Since rays $F_{em}E$, and $F_{aph}G$ are parallel,

angle $DF_{aph}G$ = angle $DF_{em}E$

and $\dfrac{DG}{F_{aph}D} = \dfrac{DE}{F_{em}D}$

$\dfrac{DG}{DE} = \dfrac{F_{aph}D}{F_{em}D} = \dfrac{23.23}{17.05}$

Therefore, RSM $= \dfrac{23.23}{17.05}$

$= 1.36$

Optical problems in correcting aphakia with spectacles

The optical correction of aphakia has already been mentioned with respect to spectacle magnification and effective power of lenses.

All aphakic spectacle wearers have to contend with several problems due to the high refractive power of the lenses required (approx. +10.0 D or more). The special problem of correcting unilateral aphakia in the presence of a normal fellow eye is discussed later.

It has been shown earlier that the relative spectacle magnification produced by aphakic spectacle correction is approximately 1.33. This means that the image produced in the corrected aphakic eye is one third larger than the image formed in an emmetropic eye. This magnification causes the patient to misjudge distances. Objects appear to be closer to the eye than they really are because of the increased visual angle subtended at the eye (Fig. 10.21).

Fig. 10.21a A standard object subtends a larger visual angle the closer it is to the eye.

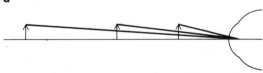

Fig. 10.21b An artificially magnified object is therefore assumed to be closer to the eye than it really is.

The image magnification also results in an enhanced performance of standard tests of visual acuity, e.g. Snellen test type. For example, a level of 6/9 for an aphakic spectacle wearer is equivalent to 6/12 for an emmetropic eye.

The use of a contact lens, or an intra-ocular implant, which reduces the RSM to 1.1 or 1.0 respectively, overcomes these problems.

The lenses used in aphakic spectacles are subject to the aberrations discussed in Chapter 8. In particular, image distortion is very troublesome to the newly aphakic patient. Straight lines appear curved except when viewed through a very small axial zone of the lens (pin cushion effect, p. 82). The linear environment thus appears as disconcerting curves, and these change their shape as the patient moves his eyes and looks through different zones of his lenses. Patients usually adapt to this by learning to restrict their gaze to the axial zone of their lenses and by moving their head rather than their eyes to look around.

The prismatic effect (Chapter 5, p. 56) of aphakic spectacle lenses produces a ring scotoma all around the edge of the lens. This scotoma may well cause patients to trip over unseen obstructions in their path.

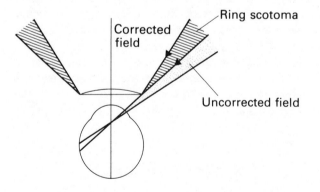

Fig. 10.22 Ring scotoma in corrected aphakia.

Furthermore, the direction of the ring scotoma changes as the patient moves his eyes and objects may appear out of the scotoma or disappear into it—the 'jack-in-the-box' phenomenon.

Fig. 10.23 Effect of eye movement on ring scotoma in corrected aphakia—Jack-in-the-box phenomenon.

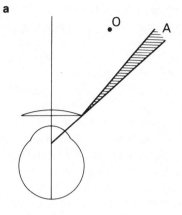

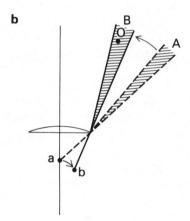

Fig. 10.23a shows the eye in the primary position and the location of the ring scotoma. An object O is visible to the patient through the periphery of the spectacle lens. Let us now consider what happens when the patient tries to look directly at object O.

Reference to Fig. 10.23b shows that as the eye rotates, moving the nodal point from a to b, the ring scotoma moves in the opposite direction, from A to B. Thus when the patient tries to look at object O it disappears in the ring scotoma B only to reappear in his peripheral vision when he looks away. This disappearance and re-emergence of an object is known as the 'jack-in-the-box' phenomenon.

Finally, high powered glass lenses are heavy and may consequently cause the spectacles to slip down the patient's nose thus altering the effective power of the lenses. (Heavy spectacles are also uncomfortable to wear.) Weight may be reduced by the use of plastic lenses, but these tend to scratch especially if laid 'face' downwards when not in use. Another means of reducing lens weight and thickness is by the use of lenticular form lenses. A lenticular lens has only a central portion or aperture worked to prescription, the surrounding margin of the lens acting only as a carrier. However, the field of vision is necessarily reduced.

The fore-going aberrations and the prismatic effect and its consequences can all be eliminated by the use of contact lenses or intra-ocular implants. The advantages of these two forms of correction stem from the fact that in each case the correcting lens becomes an integral part of the optical system of the eye.

Correction of unilateral aphakia in the presence of a normal fellow eye

Spectacle correction of a unilateral aphakic eye can achieve a clear retinal image, but with an RSM of 1.33 the image in the aphakic eye is one third larger than the image in the normal fellow eye (Fig. 10.24). The patient is unable to fuse images of such unequal size (*aniseikonia*) and complains of seeing double.

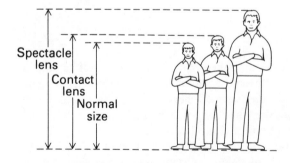

Fig. 10.24 Relative spectacle magnification in corrected aphakia.

The use of a contact lens or intra-ocular implant reduces the RSM to 1.1 or 1.0 respectively. The image in the aphakic eye will thus be the same size (intra-ocular implant) or only one tenth larger (contact lens). In either event the images can be fused and binocularity restored.

It is possible to produce a spectacle lens which has no focusing power but which alters retinal image size by increasing the visual angle subtended by an object at the eye (angular magnification, Chapter 5). Such a lens is called an *iseikonic lens*. The magnification depends on the curvature of the front surface of the lens and on its thickness. Thus, to produce pure magnification, the front curvature and lens thickness are calculated and the back curvature adjusted to render the lens afocal. Alternatively, a refractive correction may be imposed on such a lens. However, the maximum magnification which can be achieved in practice by an iseikonic lens is only about 5%, which is insufficient to be of practical benefit in the correction of unilateral aphakia.

Iseikonic lenses have been used to treat other causes of aniseikonia (different retinal image size in the two eyes), but they have fallen into disuse for several reasons. They are expensive, requiring to be individually made on special machinery. Also, they are thick and heavy to wear.

Correction of aphakia with an intra-ocular lens

The insertion of an intra-ocular lens (IOL) within the aphakic eye overcomes the optical disadvantages of aphakic spectacles and the handling and wearing difficulties encountered with contact lenses. The IOL becomes part of the optical system of the eye, and because it is situated at or very close to the position of the crystalline lens, problems with RSM do not arise.

Preoperatively, it is desirable to predict the power of the IOL which will render the individual patient emmetropic or, in some cases, produce a desired refractive error.

Many *theoretical formulae* have been devised for predicting IOL power, based on the calculation of the vergence power required in the plane of the IOL at a known position within the eye.

In the following reasoning

a = axial length of globe in metres
k = pseudophakic anterior chamber depth in metres
P_c = refractive power of cornea in dioptres
n = index of refraction of aqueous and vitreous

At the IOL, light must have a vergence equal to the dioptric value of the distance (a–k) between the IOL and the retina if the focus is to fall on the retina. Because the

IOL is not in air the distance (a–k) must be divided by n, the refractive index of the aqueous and vitreous. The dioptric value of this therefore equals $\dfrac{n}{a-k}$.

The vergence power in the plane of the IOL will be the combined effect of the refractive power of the IOL and the cornea. The effective power of the cornea in the plane of the IOL is calculated using the effective power of lenses formula, $F_2 = \dfrac{F_1}{1-dF_1}$

Substituting P_c for F_1, and $\dfrac{k}{n}$ for d (because the distance k is in aqueous of refractive index n), the effective power of the cornea in the plane of the IOL will be $\dfrac{P_c}{1-\dfrac{P_c.k}{n}}$

Therefore,

required IOL power = required vergence   –   effective power of cornea
in plane of IOL   in plane of IOL

$$= \frac{n}{a-k} - \frac{P_c}{1-\dfrac{P_c.k}{n}}$$

A more pragmatic approach to the problem of predicting IOL power is to look at the refractive results obtained in clinical practice and work back from them to a formula.

Analysis of data obtained from such retrospective study of large numbers of eyes in which IOLs have been implanted has resulted in the *Regression Analysis Intraocular Lens Formula*. In 1980 Sanders, Retzlaff and Kraff further simplified this formula producing what has become known as the *SRK formula*.

The SRK formula states that

$P = A - B(AL) - C(K)$

where   P   is the IOL power in dioptres
        A   is a constant associated with the model of IOL in use
        B   is the multiplication constant for the axial length
       AL   is the axial length in mm
        C   is the multiplication constant for the average keratometry reading
        K   is the average keratometry reading in dioptres

and the values of the multiplication constants are B = 2.5 and C = 0.9.

Substituting,

P = A − (2.5 × axial length) − (0.9 × average keratometry)
in mm   in dioptres

If a refractive condition (R) other than emmetropia is desired, the formula is modified to become

P = A − B(AL) − C(K) − D(R).

where D is the multiplication constant for the desired refraction. The value for D is 1.25 if the IOL power for emmetropia is greater than 14D, and 1.0 if the IOL power for emmetropia is less than or equal to 14D.

The SRK formula is accurate over a wide range of axial lengths, is easy to use and is widely used in clinical practice. In order to apply it, the axial length and keratometry of the individual eye must be known. The axial length is measured by A-scan ultrasonography while the corneal curvature and hence its refractive power is measured using a keratometer (Chapter 13).

Refractive state after removal of the crystalline lens from a highly myopic eye

The patient with a highly myopic eye who undergoes cataract extraction represents a special case.

The removal of the converging power of the crystalline lens reduces the total refractive power of the eye. If the degree of axial myopia is equal and opposite to the effective power of the removed lens, the eye is rendered emmetropic and the patient can see distant objects clearly without spectacles (see Fig. 10.25). This occurs after cataract extraction in axial myopia of −18 D to −20 D.

It should be recalled that the effective power of the crystalline lens *in situ* is only +15 D. It might be supposed therefore that axial myopia of −15 D would be corrected by removing the crystalline lens. But because of the change in effectivity of lenses, in order to achieve a correction of −15 D in the plane of the crystalline lens, a spectacle correction of −15 D to −20 D is required. (The exact requirement depends upon the individual characteristic of the eye in question—cf. variable states of emmetropia

Fig. 10.25

a Normal eye   **b** Axial myopia

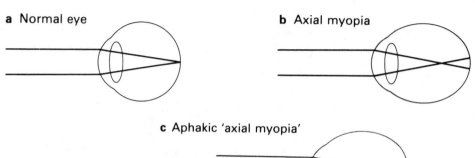

c Aphakic 'axial myopia'

Chapter 9.) Because the degree of myopia is referred to by the strength of spectacle correction it is the myope of –18 D to –20 D who becomes emmetropic after cataract extraction.

It will be recalled that the actual power of the crystalline lens in isolation is +19 D although it only contributes +15 D to the overall refractive power of the eye (p. 88). Standard intra-ocular implant lenses are therefore usually of approximately +19 D power.

Chapter 11   **Presbyopia**

The amplitude of accommodation declines steadily with age. This is due mainly to sclerosis of the fibres of the crystalline lens and changes in its capsule which reduce the spontaneous steepening of its surfaces when the ciliary muscle contracts. Also it may be that the ciliary muscle itself becomes less efficient with advancing age (after 40 years).

In infancy the eye is capable of 14 D of accommodation, but by the age of 45 years this has fallen to about 4 D. After the age of 60 years only 1 D or less remains, and part of this is probably due to depth of field, which may be enhanced by senile miosis. (A patient with no accommodation will have 0.25 D depth of field, enabling him to see clearly from 4 metres to infinity.)

In order to focus on an object at a reading distance of 25 cm, the emmetropic eye must accommodate by 4 D (see Chapter 9). However, for comfortable near vision one-third of the available accommodation must be kept in reserve. Therefore, the patient will begin to experience difficulty or discomfort for near vision at 25 cm when his accommodation has decayed to 6 D. This usually occurs between 40 and 45 years of age. A person experiencing such difficulty and discomfort for near vision due to reduced amplitude of accommodation is said to be *presbyopic*. A supplementary convex lens is used to enable the patient to achieve comfortable near vision. The lens is called a *presbyopic correction* and the inadequacy of accommodation is called *presbyopia*.

Presbyopia cannot be defined in terms of remaining amplitude of accommodation because the onset of symptoms varies with the patient's preferred working distance, the nature of the close work and the length of time for which it is done.

The amount of presbyopic correction necessary for a given patient can be calculated if the remaining amplitude of accommodation is determined (from his near point) and the desired working distance is specified. For example, an emmetropic patient has a remaining amplitude of accommodation of 3 D (near point 33 cm). In order to achieve comfortable near vision he must keep one third of this in reserve, therefore, he must use only 2 D of his 3 D of accommodation. If he wishes to see clearly at 25 cm he needs 4 D of accommodation. Thus he requires a presbyopic correction of 2 D.

In practice the refractionist learns by experience to anticipate the approximate presbyopic correction from the patient's age (see Chapter 14, p. 178). This is then confirmed by subjective refraction. In ametropia the presbyopic correction is added to the patient's distance correction.

The onset of presbyopia occurs earlier in uncorrected hypermetropia than in emmetropia, because the patient with hypermetropia must accommodate more to achieve near vision. For example, a patient with 3 D of hypermetropia needs to exert 3 D of accommodation to see clearly at infinity. Therefore, to see clearly at 25 cm 7 D of accommodation are needed (3 D + 4 D) (Fig. 11.1). Conversely a patient with 3 D of myopia has a far point at 33 cm. Thus to focus at 25 cm only 1D of accommodation is used.

Fig. 11.2 relates the three refractive states shown in Fig. 11.1 to the decline of amplitude of accommodation with age. It is apparent that the onset of presbyopia occurs earlier in hypermetropia than in emmetropia and in myopia it is delayed. Furthermore, in myopia of 4 D or more the patient can always read without glasses. However, many myopic patients prefer to use bifocal or multifocal spectacles for convenience, and to overcome any astigmatism or anisometropia that may be present.

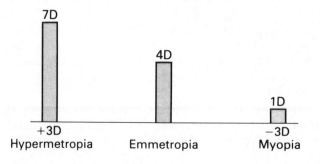

Fig. 11.1 Amplitude of accommodation necessary to achieve clear vision at 25 cm in different refractive states.

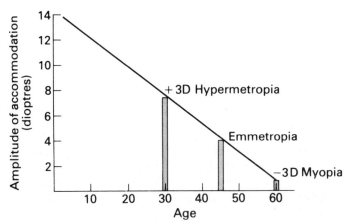

Fig. 11.2 Decline in amplitude of accommodation with increasing age.

The presbyopic correction must be adjusted for different working distances. A patient may require mid-distance glasses, e.g. for reading music, as well as for reading text.

In practice it is easy to prescribe too strong a presbyopic correction for a given task because the patient, away from his usual surroundings and anxious to perform well, tends to hold the reading test type closer to his eyes than usual. A useful safeguard against over-correction is to ensure that the patient can read N5 at his approximate reading distance but also N8 at arm's length with the proposed correction.

MULTIFOCAL LENSES

It is often inconvenient for the presbyopic patient to have separate pairs of distance and reading spectacles, and there is considerable demand for single lenses which incorporate both the distance and near correction.

Bifocal lenses

The simplest multifocal lenses are bifocal lenses, which have two portions, the distance and near portions. There are several types. The earliest form was the Franklin bifocal lens in which the two portions were cut from separate lenses and held together by the spectacle frame Fig. 11.3.

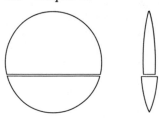

Fig. 11.3   Franklin bifocal lens.

The Franklin 'split' bifocal is seldom used today, having been superseded by the cemented wafer, the fused and the solid bifocal. The cemented wafer bifocal consists of the distance lens with a supplementary wafer cemented onto it with Canada balsam. One surface of the wafer is ground to fit the surface of the distance lens while the other supplies the extra power required for the near portion (Fig. 11.4). The wafer is known as the *segment* while the combined lens formed by the wafer and the underlying part of the distance lens is called the *near portion* of the whole lens.

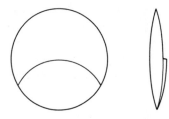

Fig. 11.4   Cemented wafer bifocal lens.

Fused bifocal lenses (Fig. 11.5) are made by combining glass of different refractive indices. The spherical surface of a crown glass lens is hollowed out to receive a segment of high refractive index flint glass and the two are fused together with heat (600°C). Fused bifocals are optically very good except that chromatic aberration may be troublesome with high reading additions (due to the different dispersions of flint and crown glass).

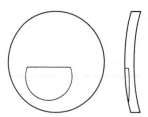

Fig. 11.5   Fused bifocal lens.

Solid bifocals (Fig. 11.6) are made from one piece of glass or plastic. The near segment is made by grinding different curvatures on the distance and near portions of the lens. The near segment is ground on to the spherical surface of a toric lens.

Fig. 11.6 Solid bifocal (executive).

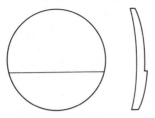

Most bifocals in current use are of the solid or fused type.

The shape, size and position of the near portion can be varied according to the occupational and cosmetic demands of the individual patient.

The design of a bifocal lens is, however, not entirely straightforward.

The points at which the visual axis of the eye passes through the spectacle lens in distance and near vision are called the Distance Visual Point (DVP) and the Near Visual Point (NVP) respectively. The NVP for reading lies 2 mm nasal and 8 mm below the DVP (Fig. 11.7).

Fig. 11.7 Bifocal lens— position of the near and distance visual points.

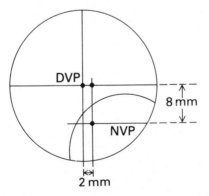

If the optical centre of each portion of the lens coincides with the NVP or DVP it is inevitable that there will be a prismatic effect at the junction of the near and distance portions of the lens. This is due to the prismatic effect at any non-axial point on a lens (see p. 56. Decentration of lenses). At the interface where the distance and near portions join, the prism power of the lens will suddenly change (the prism power depending upon the dioptric power of each portion of the lens and the distance of the interface from its optical centre). The only way to reduce this 'prismatic jump' is to make a lens in which the optical centres of the near and distance portions lie at or near the junction of the two segments. There will be no prismatic jump if the centres coincide at the interface, a mono-centric bifocal. This, however, entails moving the optical centre of each segment away from the corresponding visual point. Alternatively, prismatic jump can be reduced by incorporating a base-up prism in the reading segment. The problem of prismatic jump is most troublesome in high-powered lenses and bifocals may not be tolerated by patients with high refractive errors or with significant anisometropia. Also, great care must be exercised when prescribing for patients with extra-ocular muscle imbalance.

Another optical drawback of bifocal lenses arises from the fact that the near visual axis NR does not lie along the optical axis (NC^1C^3) of the near portion of the lens (Fig. 11.8). The astigmatic aberrations which occur when light passes obliquely through a lens (see pp. 80–1) are therefore present to some degree when the near portion is used.

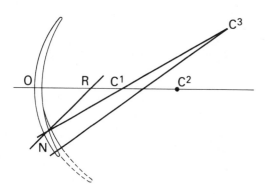

Fig. 11.8 Bifocal lens. This shows the visual axes and optical axes of the lens.

O Distance Visual Point
N Near Visual Point
R Centre of rotation of the eye
C^1 Centre of front surface of lens
C^2 Centre of back surface of lens
C^3 Centre of near segment back surface

When prescribing bifocals it is important to consider the needs of the individual patient. A typist or supermarket cashier needs to view a relatively large area at a slightly greater distance than the usual reading distance. The near segment should therefore be large and carry a slightly reduced presbyopic correction. On the other hand, the outdoor person who does little near work but needs a wide field of distance vision, benefits from having a relatively small near segment.

Certain occupations, such as working at heights, contraindicate the use of bifocals. Elderly folk who are unsteady on their feet or suffer from vertigo are also unsuitable for bifocals for they have a distorted, blurred and prismatically displaced view of the ground through the near portion.

Emmetropic patients are often well-suited by half glasses comprising a presbyopic correction alone. Conversely, the low myope may choose a spectacle lens with the lower portion cut away, enabling him to read with the naked eye.

Trifocal lenses

Trifocal lenses incorporating a further segment of intermediate power for mid-distance vision are also available (Fig. 11.9). These are of use to patients with little or no remaining accommodation (and therefore a high reading addition) for whom intermediate distance vision is blurred through either the near or distance correction.

Fig. 11.9 Trifocal lenses.

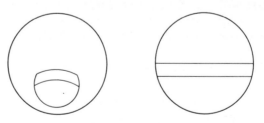

The mid-distance portion, above the near portion, usually carries half the full reading addition.

Progressive power lenses

Lenses are made in which the lens power gradually changes from top to bottom. These lenses, which are called progressive power lenses, do not look like multifocal lenses as there is no visible interface between the distance and near portions. Also the optical problems associated with a sharp interface between the distance and near portions are avoided. However, progressive power lenses have their own optical problems as described later.

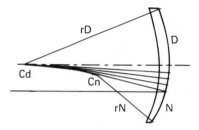

Fig. 11.10   Cross section and radii of curvature of a progressive power lens.

Fig. 11.10 shows the cross section and radii of curvature of a progressive power lens. The upper, distance, portion of the lens (D) has a constant radius of curvature, rD, while the bottom, reading, zone (N) has a constant radius of curvature, rN. Between these two portions there is a transitional zone, the radius of curvature of which gradually decreases from rD to rN. The progressive change of curvature is worked on the convex surface of the lens and is only optically 'true' along a central corridor of the lens surface, linking the distance and near portions (Fig. 11.11). This is because it is mathematically and physically impossible to design a lens without peripheral distortions and aberrations in the zone on each side of the corridor of progressive power.

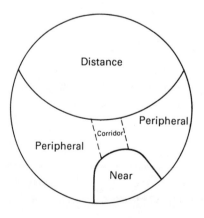

Fig. 11.11   Progressive power lens.

Many patients find the distortions intolerable, especially those individuals with a large cylinder in their prescription or a high reading addition which gives a large change in power and curvature between the distance and near portions of the lens.

Each manufacturer has his own lens design and trade name. Some progressive power lenses only achieve the full reading addition near the lower edge of the lens. If such a lens is to be dispensed, it is wise to specify an extra +0.50 DS reading addition to ensure that sufficient reading power is present at the near visual point.

Optics of Low Vision Aids Chapter 12

Magnifying devices of several kinds are in wide use to assist the poorly sighted patient in his daily life. Most so-called 'low vision aids' are designed to be used as reading aids, that is, for near vision. However, other devices exist to assist distance vision, e.g. to enable the patient to read bus numbers and watch television. In this chapter, the underlying optical principles of low vision aids are described.

For a further account of the different forms available and their clinical application the reader is recommended to Chapter 26, Duke-Elder's *Practice of Refraction* (9th Edition).

All low vision aids work by presenting the patient with a magnified view of the object. Most are optical systems which act by increasing the angle subtended by the object at the eye, thus producing an enlarged retinal image (cf. angular magnification, p. 53. The magnifying power (MP) of such an optical system can be defined as

$$MP = \frac{\text{retinal image size with use of instrument}}{\text{retinal image size without use of instrument}}$$

Projection systems are also coming into use as low vision aids—an enlarged image of the object is presented to the patient on a screen which he can view from a convenient distance. Closed circuit television is one means of achieving this and is in limited use.

The basic optics of the optical systems used as low vision aids (LVA) is described below.

The convex lens

The use of the convex lens as a magnifying loupe was described in Chapter 5. A high-power simple convex lens mounted in a spectacle frame works in the same way. The magnifying power is achieved by allowing the eye to view the object at closer range than would be possible unaided or with a standard presbyopic reading correction. However, the power of lens used e.g. +5.0 DS is less than that of the standard loupe (×8 = +32 DS) so that in this case the convex lens may be regarded as an enhanced presbyopic correction (Fig. 12.1).

a Object viewed at 40 cm with standard +2.50 DS presbyopic correction

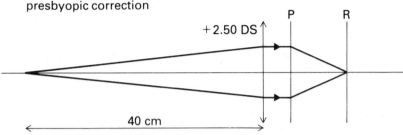

b Object viewed at 20 cm with +5.0 DS convex lens

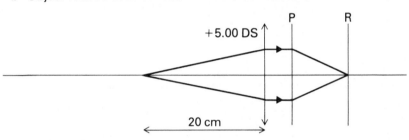

c Angle subtended at eye by object at 40 cm compared with angle subtended at eye by object at 20 cm

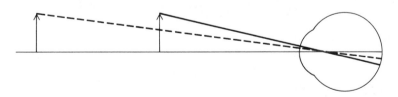

Fig. 12.1 Convex spectacle lens as LVA. Example: a 65-year old emmetrope with senile macular degeneration. (P = principal plane, R = retina).

Convex lenses are also used as hand-held magnifiers, or mounted on legs as 'stand magnifiers'. The object is located between the first principal focus and the lens and a magnified virtual image is produced which is viewed by the eye.

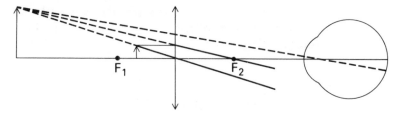

As the object moves nearer to the first principal focus, F_1, the virtual image becomes larger and is situated further from the eye. Thus a hand-held magnifier can be positioned by the user so that the image is formed at a comfortable viewing distance from the eye. Stand magnifiers are made so that the optimum object–lens distance is maintained. A variant is the 'paper weight' plano–convex lens which is very thick and rests directly on the page.

The field of vision obtained with a convex lens used as a hand or stand magnifier is dependent upon the size or aperture of the lens, and on the eye–lens distance. The greater the eye-lens distance, the smaller the field of vision.

Convex cylindrical lenses are also employed as reading aids. The bar-shaped lens which has no refractive power or only a low converging power in its long axis and high converging power in cross-section is laid on a line of print and produces vertical magnification of the letters.

Fig. 12.2   Convex lens used as magnifier when object is located between principal focus and lens.

Optics is **not really difficult**

Fig. 12.3   Convex cylindrical magnifying lens.

The Galilean System

The Galilean telescope is composed of a convex objective and a concave eye-piece lens, separated by the difference of their focal lengths. It produces an erect magnified image and the image is not greatly distorted by curvature of field or astigmatism (cf. Chapter 8). Furthermore, the system is compact and light and thus very suitable for use as a magnifying aid—usually mounted in a spectacle frame. It can be adapted for viewing near or distant objects but it is difficult to combine near and distance use in the same instrument.

Fig. 12.4 Galilean telescope.

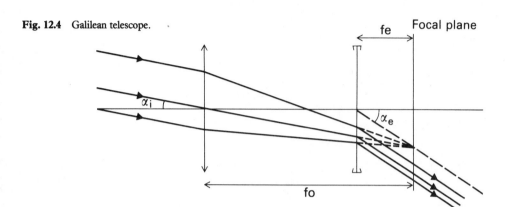

The Galilean telescope magnifies by increasing the angle subtended by the object at the eye.

Angular magnification, $M = \dfrac{\alpha_e}{\alpha_i}$

where α_e is the angle of emergence
α_i is the angle of incidence.

It can be shown mathematically that

$$M = \frac{Fe}{Fo}$$

where Fe is the power of the eye-piece lens in dioptres
Fo is the power of the objective lens in dioptres.

The practical usefulness of optical magnifying devices as low vision aids is limited by the following factors.
1. High magnification results in reduced field of view, which makes rapid scanning of a line or page of print impossible. This factor also limits the usefulness of a distance low vision aid.
2. The object to be viewed has to be held close to the eye.
3. Magnification means that depth of focus is reduced. Thus the object–lens distance is critical.
In practice, any unsteadiness of hand or head leads to unpleasant instability of field and focus.

Therefore, the aim of the prescriber should be to provide the lowest magnification that is adequate for the patient's requirements in a useable and acceptable form.

Chapter 13   **Instruments**

DIRECT OPHTHALMOSCOPE

The direct ophthalmoscope is the instrument most commonly used for routine examination of the fundus of the eye. It is small, easily portable and can also be used to examine the more anterior parts of the eye. It is important, therefore, to understand how it works, its advantages and limitations.

Fig. 13.1 The direct ophthalmoscope.

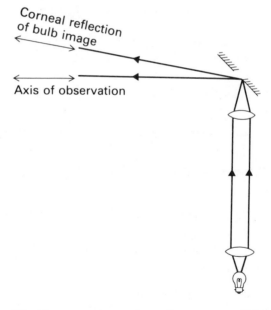

The instrument consists of a system of lenses which focus light from an electric bulb on to a mirror where a real image of the bulb filament is formed. The mirror reflects the emitted light in a diverging beam which is used to illuminate the patient's eye. The mirror contains a hole through which the observer views the illuminated eye. The image of the bulb is formed just below the hole so that its corneal reflection does not lie in the visual axis of the observer (Fig. 13.1).

The area of retina which can be seen at any one time

is called the field of view. It is governed by the projected image of the sight-hole on the retina (the sight-hole being the hole in the mirror or the observer's pupil, whichever is the smaller). The image A'B' of the sight-hole AB is constructed (Fig. 13.2) using a ray through the nodal point, N, and a ray parallel to the visual axis which is refracted by the eye to pass through its posterior focal point.

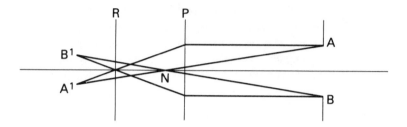

Fig. 13.3 shows the area of retina on to which image A'B' is projected, which is the field of view.

Fig. 13.2 Direct ophthalmoscope. Image of sight-hole formed by emmetropic eye.

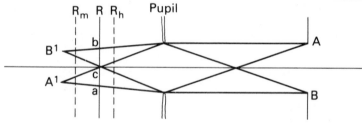

Fig. 13.3 shows that the field of view is smaller in a myopic eye, R_m, and larger in a hypermetropic eye, R_h, than in an emmetropic eye, R.

Fig. 13.3 Direct ophthalmoscope. Field of view ab (projected image of sight-hole on retina of emmetropic eye, R, myopic eye R_m and hypermetropic eye R_h).

Fig. 13.4 Direct ophthalmoscope. Effect of pupil size on field of view ab.

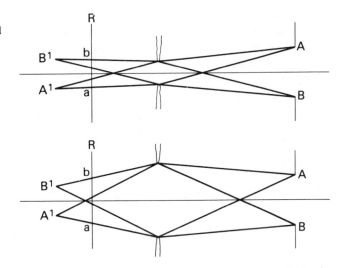

Fig. 13.5 Direct ophthalmoscope. The field of view ab increases as the sight-hole approaches the patient's eye.

Fig. 13.4 shows that the field of view is considerably enlarged when the pupil is dilated; hence the advantage of instilling a mydriatic prior to fundoscopy.

In order to utilize the maximum available field of view it is necessary for the observer to be as close as possible to the patient's eye. Fig. 13.5 shows that as the distance between the patient and the observer decreases, the field of view (the projected image of the sight-hole on the patient's retina) becomes larger.

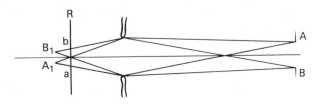

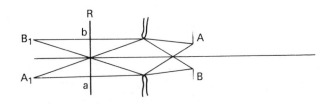

Because the real image of the bulb filament lies just below the sight-hole, the field of view is not evenly illuminated, the area bc in Fig. 13.3 being brighter than the area ac, but this effect has been largely eliminated in the most modern instruments. A dark shadow that is often seen when the peripheral parts of the retina are examined is due to total internal reflection of light at the periphery of the crystalline lens.

Light reflected from the illuminated retina of the patient's eye passes back, through the hole in the mirror and into the observer's eye. The position and size of the image formed in the observer's eye can be constructed by first constructing the image, xy, of the illuminated retina XY which is formed at the patient's far point. A ray from

Fig. 13.6   Direct ophthalmoscope. Construction of image in observer's eye.

a Emmetropic patient

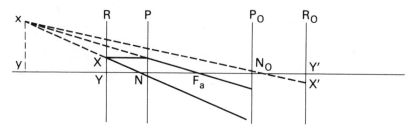

b Hypermetropic patient

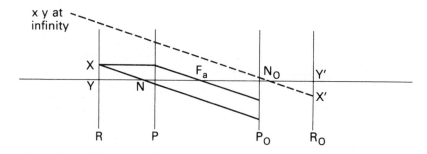

c Myopic patient

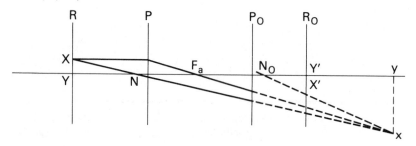

the top of that image, passing through the observer's nodal point, N_o, locates the position of the top of the image, X'Y', on the observer's retina R_o. (If this construction is used it is not necessary for the patient's and observer's anterior focal points to coincide—a condition rarely fulfilled in practice.) In the following diagrams (Figs. 13.6; 13.7; 13.8) R, P, N and F_a are the patient's retina, principal plane, nodal point and anterior focal point respectively, while R_o, P_o, and N_o refer to the observer's retina, principal plane and nodal point.

It can be seen from Fig. 13.6 that the image formed in the observer's eye is inverted and is therefore seen as erect. Also, the image size varies with the refractive state of the patient's eye, the image being smaller in hypermetropia, and larger in myopia than in emmetropia.

However, the constructions in Fig. 13.6 take no account of the ability of the observer to focus the beam of light reflected from the patient's retina on to his own retina. In the case of an emmetropic observer viewing an emmetropic patient the rays of light leaving the patient's eye are parallel and are therefore focused on the observer's retina without any accommodative effort or the use of a correcting lens (Fig. 13.7).

Fig. 13.7 Direct ophthalmoscope. Emmetropic patient and observer.

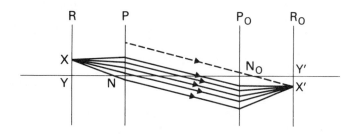

If the patient is hypermetropic, a diverging beam of light leaves his eye (Fig. 13.6b) and it behoves an emmetropic observer to accommodate or to use a correcting convex lens in order to bring the light to a focus on his retina. Fig. 13.8a is based on Fig. 13.6b, but the pencil of light around the ray which passes through N_o, the observer's nodal point, has been added to show that the point X' (and Y') in Fig. 13.6b is really a blur circle as a virtual image $x_o y_o$ is formed behind the observer's retina. The observer's view is therefore blurred.

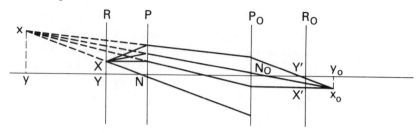

Fig. 13.8a Direct ophthalmoscope. Hypermetropic patient, emmetropic observer.

Likewise, an emmetropic observer viewing a myopic eye receives a converging beam of light which is brought to a focus in front of his retina (Fig. 13.8b). Once again, X′ is a blur circle and the observer sees a blurred image unless he uses a correcting concave lens.

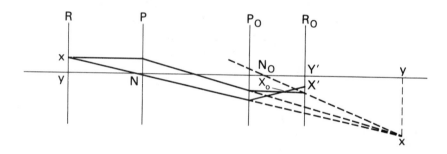

Fig. 13.8b Direct ophthalmoscope. Myopic patient, emmetropic observer.

In order to see a clear view of the patient's retina in myopia or hypermetropia it is therefore necessary to use a correcting lens to bring the image to a focus on the observer's retina. A range of correcting lenses is incorporated in the ophthalmoscope.

The correcting lens power should be equal to the degree of convergence or divergence of the light emerging from the patient's eye (cf. Vergence of Light, p. 52, Fig. 5.8). Thus, the correcting lens will render the beam parallel and the emmetropic eye of the observer will form a focused image on its retina X_rY_r. This focused image can be constructed by drawing the emerging light and image xy of XY formed at the patient's far point as before (cf. Fig.

13.6). A ray from x, passing through the centre of the correcting lens determines the direction of the parallel beam formed after refraction of the emerging beam by the correcting lens. The ray in the parallel beam which passes through the observer's nodal point N_o locates X_r on the retina (Fig. 13.9).

(a) Emmetropic patient

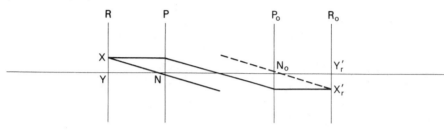

(b) Hypermetropic patient

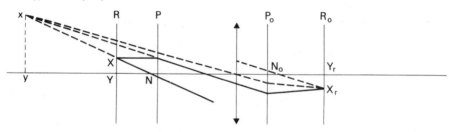

(c) Myopic patient

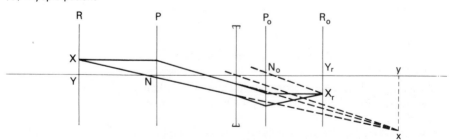

Fig. 13.9 Direct ophthalmoscope. Construction of image in observer's eye when a correcting lens is used.

The image formed on the observer's retina is smaller when a hypermetropic eye is viewed and larger when a myopic eye is viewed than when an emmetropic eye is examined, but, the use of a correcting lens reduces the discrepancy in size (Fig. 13.9). (If the patient's and the observer's anterior focal points coincide, and the correcting lens is placed at that point, the observer's image is the same size regardless of whether the patient is hypermetropic, emmetropic or myopic. However, these conditions are rarely fulfilled in practice.)

The smaller image size when a hypermetropic eye is examined accounts for the relatively small image seen in aphakia as compared with that in emmetropia. Referring back to Fig. 13.3 reminds us that the field of view is wider in hypermetropia. Thus, when examining very hypermetropic eyes a small image of a wide field of view is seen and the whole fundus can be scanned quickly.

The enlargement of image size seen when a myopic eye is examined, coupled with the reduced field of view as compared with an emmetropic eye (Fig. 13.3) results in the observer seeing a magnified but restricted view of the myopic fundus. Also, in axial myopia the eye itself is bigger than an emmetropic eye. Thus it is difficult to examine a myopic fundus using the direct ophthalmoscope because the field of view is so small compared to the size of the fundus.

The view of the fundus when seen through the direct ophthalmoscope in high degrees of hypermetropia or myopia is so different and characteristic that the refractive state of the eye is betrayed to the observer. This disparity of appearance can be made use of when the refractionist cannot interpret the dull and indistinct reflex seen during retinoscopy in high degrees of ametropia.

It is impossible to secure a perfect view of the fundus of an astigmatic eye because the only correcting lenses in the ophthalmoscope are spherical. It is thus only possible to correct one meridian at a time. If the degree of astigmatism is high, the difference in image size due to the disparity of dioptric power of the eye in the two principal meridians causes distortion of the image and the optic disc appears oval.

Thus far it has been assumed that the observer is emmetropic. For those observers who have a refractive error there are two possibilities. The observer can remove his spectacles and rack up the appropriate lens in the ophthalmoscope to give a clear view of the patient's fundus. The appropriate lens is the algebraic sum of his own and the patient's refractive error (the value is approximate because the position of the correcting lens influences its effectivity [see p. 102]). Alternatively, he can use the instrument with his glasses on. However, his field of view will be restricted as the sight hole in the mirror will be further from his eye.

The patient's refractive error can be roughly judged by noting the power of the correcting lens used. This, however, assumes that neither patient nor observer is accommodating.

The posterior pole of a highly myopic fundus is best seen with the direct ophthalmoscope if the patient keeps his glasses on. The magnification of the patient's retina when viewed through the direct ophthalmoscope may be calculated. The underlying principle is the same as that of the loupe (Chapter 5). The observer is using the dioptric power of the patient's eye as a loupe and is thus able to inspect the patient's retina at close quarters, i.e. well within his near point of distinct vision, and yet see it clearly.

The formula for magnification achieved by a loupe is $M = \frac{F}{4}$ where M is the magnification and F the dioptric power of the loupe. If we ascribe a dioptric power of +60 D to the patient's emmetropic eye, the magnification of the direct ophthalmoscope is ×15. This degree of magnification makes the direct ophthalmoscope particularly useful when examining patients with retinopathy, for it allows most micro-aneurysms to be seen. However, some are too small to be visualized with this instrument. Many modern direct ophthalmoscopes incorporate a red-free filter. The resulting green light causes the micro-aneurysms to show up as black dots against a green background and this makes their detection easier.

The direct ophthalmoscope can also be used as a self-illuminating loupe to examine the anterior segment of the eye, e.g. lens opacities can be directly inspected through the +10 D correcting lens.

INDIRECT OPHTHALMOSCOPE

The indirect ophthalmoscope offers an alternative means of examining the retina and vitreous, and gives a very different view of the retina compared with that obtained with the direct ophthalmoscope. Both instruments have their place and should be used to complement each other during the clinical examination of the eye. A table setting out the characteristics of the two instruments can be found on p. 147.

In the indirect method of ophthalmoscopy a powerful convex lens (hereafter called the condensing lens) is held in front of the patient's eye. The usual powers used are +20 D and +13 D. The illuminating light beam passes through the condensing lens into the eye and light reflected from the retina is refracted by the condensing lens to form a real image between the condensing lens and the observer. The observer studies this real image of the patient's retina, Fig. 13.10.

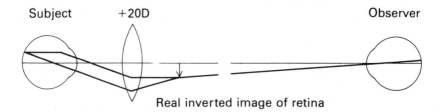

Subject   +20D   Observer

Real inverted image of retina

Illumination is usually provided by an electric lamp mounted on the observer's head. Light from this source is rendered convergent by the condensing lens. Thus a convergent beam enters the patient's eye and is brought to a focus within the vitreous by the eye's refractive system. The light then diverges again to strike the retina (Fig. 13.11). The illumination is therefore bright and even, as it comes from the real image of the light source within the patient's eye.

Fig. 13.10 Indirect opthalmoscope.

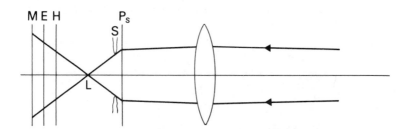

M E H   P$_s$
                 S
                 L

Fig. 13.11 illustrates the path of the illuminating beam, P$_s$ being the subject's principal plane, S his pupil and M, E and H being the retina in myopia, emmetropia and hypermetropia respectively. The diagram shows that the field of illumination is largest in myopia and smallest in

Fig. 13.11 Indirect ophthalmoscope. Field of illumination.

hypermetropia and that in all refractive states the size of the subject's pupil limits the field of illumination.

The condensing lens is held in front of the patient's eye at such a distance that the patient's pupil and the observer's pupil are conjugate foci. (This means that light arising from a point in the subject's pupillary plane is brought to a focus by the condensing lens in the observer's pupillary plane, and vice versa.) A reduced image of the observer's pupil is therefore formed in the subject's pupillary plane (Fig. 13.12) (the image of a 4 mm pupil is approximately 0.7 mm).

Fig. 13.12 Indirect ophthalmoscope. Construction of the reduced image O_1 of the observer's pupil O in the subject's pupillary plane S.

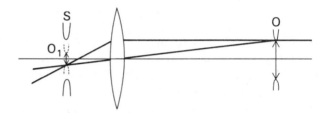

Only those rays of light which leave the subject's eye via the area of the image of the observer's pupil can, after refraction by the condensing lens, enter the observer's pupil and be seen by him. The observer's pupil is the 'sight hole' of the system and its size influences the field of view.

The field of view is also limited by the aperture or size of the condensing lens. Only those rays which leave the subject's eye via the image of the observer's pupil and which then pass through the condensing lens will be seen by the observer, Fig. 13.13.

Fig. 13.13 Indirect ophthalmoscope. The field of view is limited by the image of the observer's pupil O_1 in the subject's pupil S and by the aperture of the condensing lens. (a) When a large aperture condensing lens is used the field of view is limited only by the observer's pupil O_1. (b) When a small aperture condensing lens is used it is the lens aperture that limits the field of view.

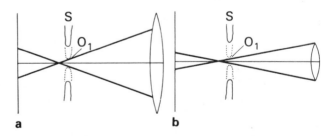

Thus, to recap, the subject's pupil size determines the field of illumination, Fig. 13.11. The observer's pupil size and the aperture of the condensing lens determine the field of view, Fig. 13.13a and b.

It is usual to dilate the patient's pupil widely prior to

indirect ophthalmoscopic examination, in order to widen the field of illumination. It is not practical to dilate the observer's pupil because his visual acuity would be impaired. This is because he would suffer an increase in aberrations and loss of accommodation, the latter being particularly troublesome for the low hypermetrope. Therefore a condensing lens is chosen with the largest possible aperture to give the widest field of view. Condensing lenses of wide aperture must be of aspheric form to minimize aberrations. Using such a lens, a field of view of approximately 25° can be achieved, i.e. four times larger than the field of view of the direct ophthalmoscope. This is a great advantage when a highly myopic eye is to be examined.

Light emerging from the patient's eye is refracted by the condensing lens to form a real image of the retina between the condensing lens and the observer. The image is both vertically and laterally inverted, (upside down and back to front). It is situated at or near the second principal focus of the condensing lens, i.e. approximately 8 cm in front of a +13 D lens, Fig. 13.14. The observer holds the condensing lens at arm's length and thus views the image from a distance of 40–50 cm. Therefore, to see the image clearly the observer must accommodate or use a presbyopic correction. The binocular indirect ophthalmoscopes have +2.0 D lenses incorporated in the binocular prismatic eyepieces so that the observer does not need to accommodate. (However, those observers whose near correction is more than +2.0 DS because of underlying hypermetropia, or who have any significant refractive error, need to wear their spectacle correction when using the instrument).

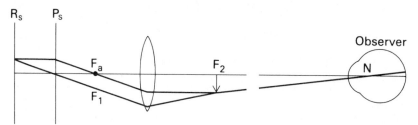

The linear magnification of the image can be calculated from Fig. 13.15.

Parallel light emerges from the retina AB of the emmetropic subject's eye, and is refracted by the condensing lens to form an image ab in its principal focal plane F.

Fig. 13.14 Indirect ophthalmoscope. Formation of the real image of the subject's retina. The image is viewed by the observer (image-observer distance foreshortened).

Fig. 13.15 Indirect ophthalmoscope. Linear magnification.

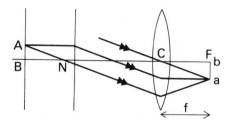

Linear magnification $= \dfrac{ab}{AB}$

However, angle aCF $=$ angle ANB because Ca and AN are parallel.

$$\text{Tan aCF} = \frac{ab}{CF} = \text{tan ANB} = \frac{AB}{BN}$$

Therefore $\dfrac{ab}{AB} = \dfrac{CF}{BN}$

CF is the focal length of the condensing lens.

BN is the distance between the nodal point and the retina of the subject's eye. If this distance BN is taken to be 15 mm, the linear magnification is equal to the focal length of the lens (in mm) divided by 15. Thus, the linear magnification of a +13 D lens, (f = 75 mm), is approximately $\times$ 5, while the linear magnification of a +20 D (f = 50 mm), lens is approximately $\times$ 3.

The angular magnification also can be calculated, and once again a +13 D lens magnifies approximately $\times$5 while a +20 D lens magnifies approximately $\times$ 3. The exact values depend on the distance from which the observer views the real image of the subject's retina, and upon the distance between the condensing lens and the subject's eye if this is ametropic (see below).

The refractive state of the patient's eye affects the size and position of the real image formed by the condensing lens.

The image of the retina of an emmetropic eye is always located at the second principal focus of the condensing lens, regardless of the position of the lens relative to the eye. This is because all rays emerging from an emmetropic eye are parallel (Fig. 13.15). Rays emerging from a hypermetropic eye are divergent, and the real image is therefore formed outside the second principal focus of the condensing lens (Fig. 13.16). Emergent rays from a myopic eye are convergent, and the real image is therefore always within the second focal length of the lens.

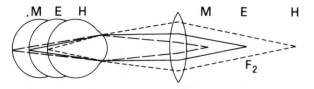

Fig. 13.16 Indirect ophthalmoscope. Relative positions of the image in hypermetropia H, emmetropia, E, and myopia, M.

Knowing the location of the image in hypermetropia and myopia, it is possible to study the effect of moving the condensing lens relative to the eye.

The following diagrams (Fig. 13.17) show the changes in image size when the first principal focus of the condensing lens F_1 is moved relative to the anterior focus Fa of the eye. A ray parallel to the optical axis of the eye which, after refraction at the principal plane, passes through Fa is used to determine the relative image sizes in hypermetropia, emmetropia and myopia.

Fig. 13.17 Indirect ophthalmoscope. To show the effect on image size of moving the condensing lens progressively further away from the subject's eye in ametropia.

a The first principal focus of the condensing lens F_1 is closer to the eye than the anterior focus of the eye F_a

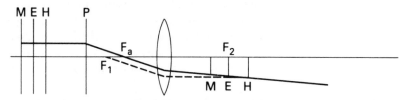

b The first principal focus of the condensing lens F_1 is at the anterior focus of the eye F_a

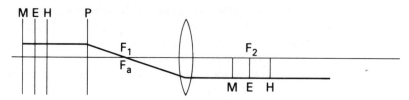

c The first principal focus of the condensing lens F_1 is beyond the anterior focus of the eye F_a

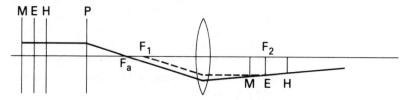

The image size in emmetropia is the same in all positions of the condensing lens, and is determined by a ray through F_1 (dotted) parallel to the emergent ray through Fa, which after refraction by the condensing lens is parallel to the principal axis of the lens.

In myopia, the image size increases as the condensing lens is moved away from the eye, while in hypermetropia the image becomes smaller. This effect can be used by the examiner to assess the refractive condition of the subject's eye.

The relative merits of the direct and indirect ophthalmoscope

Table 13.1 Summary of the optical properties of the direct and indirect ophthalmoscope

	Direct ophthalmoscope	Indirect ophthalmoscope
Image	Not inverted	Vertically and horizontally inverted.
Field of view	Small (6°)	Large (25°)
Magnification	Large (×15)	Small (×3 [+20 D]) (×5 [+13 D])
Binocularity	Not available	Stereoscopic view
Influence of patient's refractive error	Large	Small
Teaching facility	None	Teaching mirror (Fig. 3.4 and text)

Other considerations include:
1. Size of the instrument. The direct ophthalmoscope is much smaller and lighter than the indirect, and many models are pocket size.
2. Illumination is crucial to the view obtained. Because indirect ophthalmoscopes are larger instruments there is more scope for fitting more powerful light sources. This renders the indirect ophthalmoscope more useful in examining patients with opacities in the ocular media.
3. The combination of good illumination and wide field of view makes the indirect ophthalmoscope the instrument of choice when examining patients with retinal detachments. If there is extensive subretinal fluid and the

illumination is poor an underlying malignancy may not be seen with the direct ophthalmoscope. Futhermore, examination with the indirect ophthalmoscope also allows indentation of the peripheral retina.

4. The indirect ophthalmoscope is also the ophthalmoscope of choice for use during retinal detachment surgery because it is used at a distance which allows the surgeon to preserve a sterile operative field.

RETINOSCOPE

An accurate objective measurement of the refractive state of an eye can be made using the retinoscope. The technique is called *retinoscopy* or *skiascopy*.

Light is directed into the patient's eye to illuminate the retina (the illumination stage).

An image of the illuminated retina is formed at the patient's far-point (the reflex stage).

The image at the far-point is located by moving the illumination across the fundus and noting the behaviour of the luminous reflex seen by the observer in the patient's pupil (the projection stage).

Illumination stage

Figs. 13.18 and 13.19 a and b illustrate the simple system of illumination that was used before hand-held electric retinoscopes came into use. A light source was located beside the patient's head and light reflected into the patient's eye from a plane or concave mirror held by the observer. The observer viewed the patient's eye through a small hole in the mirror. The electric retinoscope has largely replaced this system. However, the principles and nomenclature have remained unchanged.

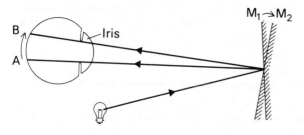

Fig. 13.18 Illumination stage. Plane mirror.

When a plane mirror is used, light is moved across the patient's fundus from A to B by rotating the plane mirror from M₁ to M₂. Note that the illuminating rays move in the same direction as the mirror.

Fig. 13.19a Illumination stage. Plane mirror. Virtual images S_1 and S_2 of light source S (located by their normals to the mirror [see Chapter 2]) throw light via the nodal point so that the movement of illumination at the retina R_1 and R_2, is 'with' the movement of the mirror, M_1 to M_2.

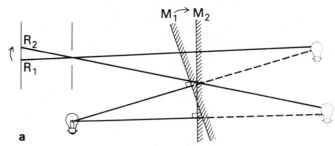

A concave mirror of focal length less than the distance between patient and observer is occasionally used for retinoscopy. A real image of the light source is formed between the patient and the observer, and close to the patient's eye. This real image acts as a bright light source for illuminating the patient's retina. However, the illumination moves in the opposite direction to the mirror (see Fig. 13.19b). Thus the movement of the reflex seen by the observer is reversed compared to the plane mirror.

Fig. 13.19b Illumination stage. Concave mirror. Real images S_1 and S_2 of the light source throw the light via the nodal point so that the illumination at the retina moves in the opposite direction or 'against' the movement of, the mirror, M_1 to M_2.

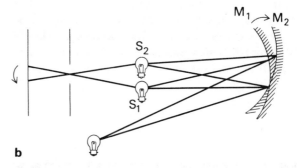

The modern electric retinoscope incorporates both these systems of illumination. This is achieved by the use of a condensing lens which can be moved within the shaft of the instrument. At its lowest position the plane mirror effect is obtained and at its highest position the effect is that of the concave mirror.

If the condensing lens is moved from the lowest (plane mirror) position upwards along the shaft of the instrument an intermediate position is reached at which a focused image of the retinoscope bulb filament falls on the patient's eye (or cheek). The effect is that of a concave mirror of focal length equalling the retinoscope–patient distance and is of no value for retinoscopy as the image of the light source is coincident with the patient's eye. However, if the retinoscope condensing lens is moved a short distance back down the shaft of the retinoscope, the 'plane mirror effect' is regained but with a much brighter illumination than that obtained with the condensing lens at its lowest position. This is because a virtual image S_1 of the light source is formed just behind the patient's eye (Fig. 13.20). The retinoscope is now acting as a concave mirror of focal length slightly exceeding the observer–patient distance.

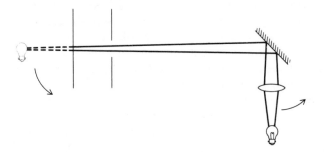

Fig. 13.20   Electric retinoscope adjusted for use as concave mirror of focal length slightly exceeding the observer–patient distance, thus obtaining *Plane Mirror Effect*. The whole instrument is tilted to achieve movement of the illumination across the patient's retina (arrows).

This arrangement is useful in clinical practice for two reasons. Firstly, the plane mirror effect (which most observers prefer) is retained. And, secondly, a bright light is achieved which makes retinoscopy easier when the pupil is small and when there are opacities in the media of the eye.

Reflex stage

Because the plane mirror effect is usually preferred in retinoscopy, the following description and diagrams illustrate the plane mirror effect.

An image A_1B_1 of the illuminated retina is formed at the patient's far-point (p. 99–100). This image may be constructed using three rays (Fig. 13.21).

1. A ray from point A of the retina R on the principal axis of the eye, which leaves the eye along the principal axis.
2. A ray from a retinal point B, off the principal axis,

which travels parallel to the principal axis as far as the principal plane, P, of the eye, where it is refracted to pass onward through the anterior principal focus, Fa, of the eye.

3. A ray from retinal point B which passes undeviated through the nodal point, N.

(NB When drawing these diagrams from memory, construct the image at the far-point for each refractive condition and Fa will look after itself. Do not put Fa at a random position and then try to make the diagram 'work').

Fig. 13.21 Reflex stage of retinoscopy.

a Hypermetropia

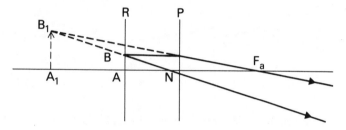

b Emmetropia

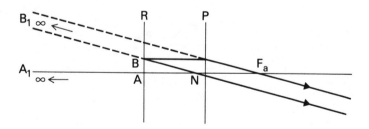

c Myopia

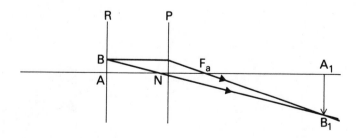

Projection stage

The observer views the image A_1B_1 of the illuminated retina AB from a convenient distance, usually ⅔ m. Fig. 13.22 depicts this and is constructed by drawing Fig. 13.21 and adding a hypothetical ray from point B_1 passing through the observer's nodal point, N_o, to the observer's retina, R_o. This ray locates the point B_o, the image of B_1 on the observer's retina and allows completion of the diagram. The observer does not see the actual image A_1B_1, but rays from A_1B_1 are seen as an illuminated area or reflex in the patient's pupil.

Fig. 13.22a Projection stage. Hypermetropia.

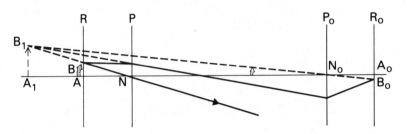

In hypermetropia the luminous reflex seen in the patient's pupil moves in the same direction as the illuminating light—a 'with' movement, indicated by the arrows in Fig. 13.22.

Fig. 13.22b Projection stage. Emmetropia.

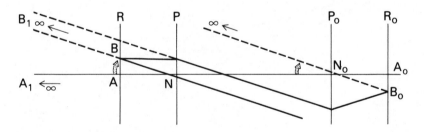

Once again a 'with' movement is observed.

Fig. 13.22c Projection stage. Myopia les than –1.5 D (for working distance of 2/3 metre).

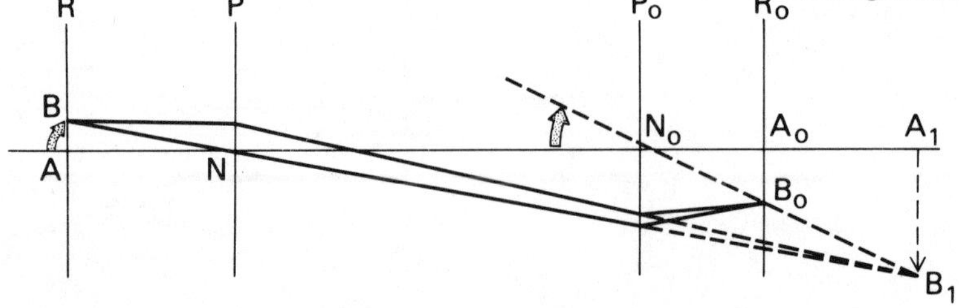

When the patient's myopia is less than the dioptric value of the observer's working distance a 'with' movement is still obtained.

The above sequence of diagrams illustrates that the angle $A_oN_oB_o$ increases progressively as a refractive error equal to the dioptric value of the observer's working distance (1.5 D) is approached. Therefore the luminous reflex appears to move more rapidly as this point is approached.

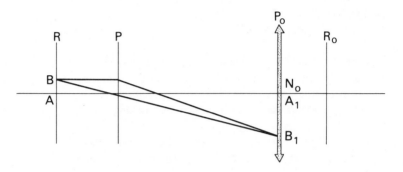

Fig. 13.22d Projection stage. Point of reversal.

The *point of reversal* or *neutral point* of retinoscopy is reached when the patient's far-point coincides with the observer's nodal point (Fig. 13.22d). No image of B_1 can be formed on the observer's retina and at this point no movement of the reflex can be discerned in the patient's pupil. The observer sees a diffuse bright red reflex. This is because the movement of the reflex is infinitely rapid.

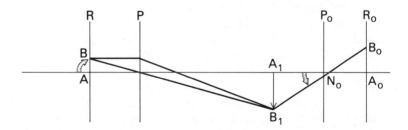

Fig. 13.22e Projection stage. Myopia greater than –1.5 D (for working distance of 2/3 metre).

When the patient's myopia exceeds the dioptric value of the working distance, the image A_1B_1 falls between the patient and the observer. The diagram shows that the luminous reflex now appears to move in the opposite direction to the illuminating light; 'against' movement. Once again the luminous reflex appears to move more rapidly as the point of reversal is approached.

In practice, lenses are placed in front of the patient's eye until the point of reversal is seen by the observer. A correction is made for the working distance (add -1.5 D for ⅔ m, add -1.0 D for 1 m) and the corrected value of the lenses equals the patient's refractive error. A full discussion of practical retinoscopy may be found in Chapter 14. Automated refraction is discussed in Chapter 15.

INSTRUMENTS USED TO STUDY CORNEAL CURVATURE

The anterior corneal surface is the main refracting surface of the eye (p. 88). Its curvature is crucial to the refracting power and optical properties of the eye. A small change in curvature or irregularity of the corneal surface has a profound effect upon visual acuity.

Accurate measurement of the corneal curvature is important in ophthalmology and indeed essential in contact lens fitting.

The anterior corneal surface reflects a small portion of any light incident upon it and thus acts as a convex mirror. The corneal surface and curvature can be examined by studying the catoptric image so formed (see p. 93).

PLACIDO'S DISC

The general shape of the cornea can be studied using Placido's disc. This is a flat disc bearing concentric black and white rings. A convex lens is mounted in an aperture in the centre of the disc. The examiner looks through the central aperture and observes the image of the disc formed by reflection from the patient's cornea. (Best results are obtained by ensuring that the disc is brightly illuminated by adjusting the light behind the patient's head leaving the patient's eye in shadow. The patient's eye can be viewed at close quarters thanks to the convex viewing lens.)

The nature of the image seen betrays the regularity or distortion of corneal curvature to the examiner.

Fig. 13.23 Placido's disc.

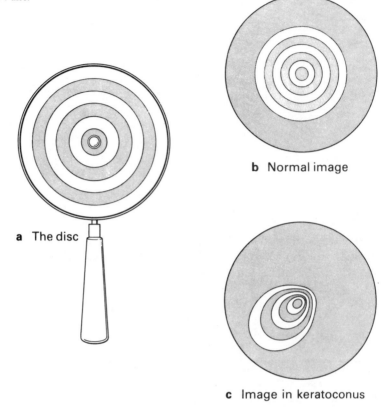

a The disc

b Normal image

c Image in keratoconus

KERATOMETER OR OPHTHALMOMETER

The radius of curvature of the central area of the cornea can be measured using a keratometer (sometimes called an ophthalmometer). Two main types are available but both measure the radius of curvature of a central zone of the cornea approximately 3 mm in diameter.

The central or axial area of the cornea, about 4 mm in diameter, is usually assumed to be a spherical refracting surface. The keratometer thus reads within this zone. The radius of curvature of the axial zone of the emmetropic eye is about 7.8 mm. The more peripheral cornea is flatter and non-spherical. Optically, it is the central 4 mm spherical zone which is utilized for vision, the periphery being screened off by the pupil. Visual acuity suffers when the pupil is widely dilated.

The principle on which the keratometer works is explained below.

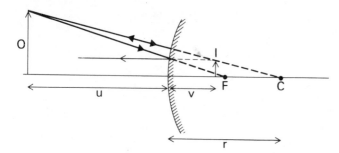

Fig. 13.24   Keratometer.
Underlying optical principle.

$$\frac{I}{O} = \frac{v}{u}$$

In practice, I is located very close to F, therefore v may be taken to equal r/2 where r is the radius of curvature of the reflecting surface. Substituting,

$$\frac{I}{O} = \frac{r}{2u}$$

$$r = 2u \times \frac{I}{O}$$

In all keratometers u is constant, being the focal distance of the viewing telescope. The von Helmholtz keratometer has a fixed object size and the image size is adjusted to measure the corneal curvature while in the Javal Schiøtz instrument the object size is varied to achieve a standard image size.

In order to study closely and measure the image formed by reflection at the cornea, some means must be adopted to overcome the natural movements of the patient's eye. Even small movements cause the image to 'dance about'.

This difficulty has been overcome by doubling the image seen by the examiner. Thus if the patient's eye moves, both images move together and a reading can be made by aligning the images one with the other.

The keratometer (ophthalmometer) of von Helmholtz

The von Helmholtz keratometer uses two rotating glass plates to achieve doubling of the image.

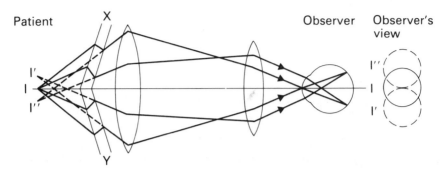

Fig. 13.25 Von Helmholtz keratometer and measurement of corneal curvature.

A beam of light which has passed through a graticule is shone on to the patient's cornea where an image I of the graticule is formed by reflection. The reflected light passes back into the instrument through two parallel-sided glass plates X and Y which are inclined to each other. These plates displace the light laterally as it passes through them, thus giving rise to two virtual images of I, I' and I'' which are viewed through a telescope. The angle of inclination of the glass plates is varied by the observer until the edges of I' and I'' touch. The distance between their centres then equals the diameter of I, from which the corneal curvature can be calculated. In fact, the instrument is calibrated in terms of corneal radius of curvature and of dioptric power of the cornea.

The instrument can be rotated to allow measurement of astigmatism in a similar manner to the Javal Schiøtz keratometer.

The Javal Schiøtz keratometer (ophthalmometer)

The Javal Schiøtz keratometer uses an object of variable size. The object consists of a pair of mires, A and B, mounted on curved side arms which project each side of the viewing telescope (Fig. 13.26).

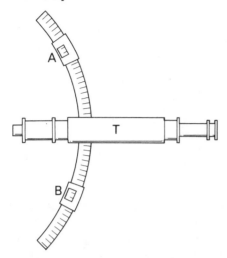

Fig. 13.26   The Javal-Schiøtz
keratometer.

Each mire consists of a small lantern with a coloured window. One mire is step shaped while the other is rectangular. The *space between the mires ab* is the object size used in the measurement (Fig. 13.27).

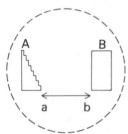

Fig. 13.27   Javal-Schiøtz keratometer. The mires.

The arms on which the mires are mounted can be rotated about the axis of the telescope so that readings can be made in any meridian.

Doubling of the image formed by reflection at the cornea is achieved by a Wollaston prism which is incorporated in the viewing telescope. A Wollaston prism consists of two rectangular quartz prisms cemented together. Quartz is a doubly refracting substance, that is, it splits a single beam of incident light to form two polarized emergent beams. By cementing two quartz prisms together with the optical 'grain' of the crystal at right angles it is possible to separate the two emergent beams by a fixed angle, while the dispersion produced by the first prism is neutralized by that of the second prism allowing sharp images to be formed (Fig. 13.28).

Fig. 13.28 Wollaston prism.

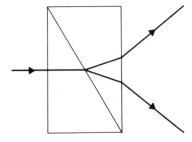

In order to measure corneal curvature the distance between the mires is adjusted until the doubled images just touch (Fig. 13.29).

Fig. 13.29 Javal-Schiøtz keratometer. Appearance of images of mires.

Object size too small

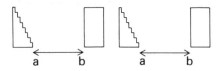

Object size correct

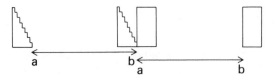

Object size too large

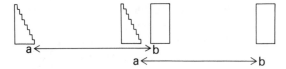

The instrument is calibrated in terms of corneal radius of curvature and in terms of dioptric refracting power of the cornea. The mires are designed so that each step of mire A (Fig. 13.27) is equivalent to one dioptre of corneal power. Thus if the inner images of a and b (Fig. 13.29) are aligned correctly in one corneal meridian but overlap by one and a half steps in the meridian at 90° to the first, 1.5 D of corneal astigmatism is present (Fig. 13.30a).

When an astigmatic cornea is examined, the two images are displaced vertically in all but the two principal meridians of the cornea. Thus the axis of astigmatism as well as its magnitude can be measured. (Fig. 13.30b). The mires of the Haag-Streit Javal Schiøtz keratometer incorporate a horizontal line to facilitate vertical alignment.

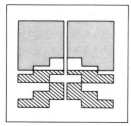

a 1½ dioptre corneal astigmatism

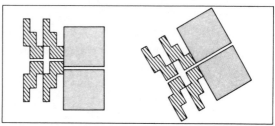

b Adjustments for reading the axis of astigmatism and corneal radius

Keratometers may also be used to measure the curvature of contact lenses.

Fig. 13.30 Javal-Schiøtz keratometer. (Haag-Streit mires) (only central pair of mires shown).

COMPOUND MICROSCOPE

A microscope provides a magnified view of a near object, as opposed to a telescope which magnifies distant objects.

The compound microscope is used in many ophthalmic instruments to provide a magnified view of the eye. The slit-lamp microscope and operating microscope are obvious examples, but the keratometer and all the instruments used in conjunction with the slit-lamp, the pachometer, applanation tonometer, gonioscopy lens, etc., depend on the microscope to provide the observer with a magnified image. The specular microscope is simply a specially modified microscope which allows examination and photography of the corneal endothelium.

The compound microscope consists of two convex lenses, the objective and eyepiece lenses (Fig. 13.31).

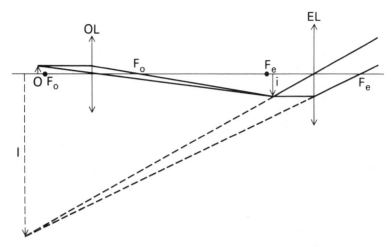

Fig. 13.31 The compound microscope.

The object O to be studied is placed just outside the anterior focal point, F_o of the objective lens OL. A real, inverted, magnified image, i, is formed some distance behind the objective lens. The eye piece lens, EL is placed so that the image formed by the objective falls at or close to its principal focal plane, F_e. The eyepiece thus acts as a loupe (see p. 54–6) and further magnifies the image seen by the observer. The final image I is vertically and horizontally inverted. Porro prisms (p. 45) are incorporated in clinical microscopes to obtain an erect non-inverted image and to shorten the physical length of the instrument.

The operating microscope and most binocular slit-lamps consist of two such compound microscopes mounted at an angle of 13° or 14° to each other, thus giving the observer a binocular, stereoscopic view. Also each lens of the microscope is replaced by a system of lenses to reduce spherical and chromatic aberration and coma.

SLIT LAMP

The routine method of examining the outer segment of the eye is with a slit lamp. It consists essentially of a relatively low-powered binocular compound microscope (see above) which is linked to an adjustable bright light source. The instrument is known as a slit lamp because in every day use the illumination is arranged so that a narrow vertical slit of light is projected on to the eye.

There are several different types of slit lamp but they all share certain basic features. The alignment of the microscope and the illumination is such that the point on which the microscope is focused corresponds to the point on which the light is focused. This is achieved by the microscope and the lighting system having a common focal plane. Their common axis of rotation also lies in this focal plane (Fig. 13.32).

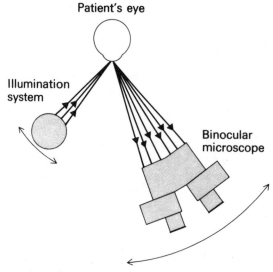

Fig. 13.32   Slit lamp.

Another common feature is that there is a considerable distance between the microscope itself and the patient's eye. This is because the microscopes have a long working distance. The resulting gap between the microscope and the patient allows the observer to carry out certain manoeuvres such as removing a foreign body from the cornea. It also gives room to interpose certain optical devices, for example, a Hruby lens or a 3-mirror contact lens, which permit inspection of the vitreous and retina. The microscope tubes themselves are shortened by the incorporation of prisms, which also invert the image vertically and horizontally so that it appears erect and the right way round to the observer.

Methods of examination

A wealth of information can be gained by examining the eye properly with a slit lamp. Certain modifications in terms of the illumination, including the use of filters, can be of considerable benefit in eliciting important physical signs.

Direct focal illumination

This is the most generally useful method of illuminating the eye. The slit beam is accurately focused upon that part of the eye under inspection.

Diffuse illumination

The beam of light is thrown slightly out of focus across the structure being examined so that a large area is diffusely illuminated. This can be a particularly helpful way of looking at the anterior capsule of the lens.

Lateral illumination

This technique involves illuminating the structure being examined by light which is reflected from tissue just to one side of it. For instance, if a beam of medium width is directed upon the margin of the pupil, the outer rim of the sphincter muscle becomes apparent.

Retro-illumination

This is a method of examining a particular part of the eye by light reflected from a structure behind it. The structure behind is used as a mirror to illuminate the part of the eye in question. A good example of retro-illumination is when areas of iris atrophy are identified by light which is reflected from the choroid. The illuminating column of the slit lamp should be brought to lie between the objective lenses of the microscope so that the illuminating and viewing systems are co-axial.

Specular reflection

A great deal of information can be gained about the nature of a mirror-like surface by examining the rays of light reflected from it. The corneal surfaces and the anterior lens capsule can be examined in this way. Bearing in mind the laws of reflection (Chapter 2) the patient's gaze is directed to bisect the angle between the axis of illumination and that of the microscope. This is the best way of inspecting the corneal endothelium with the slit lamp.

Sclerotic scatter

When the slit beam is directed on to the limbus at, for example, the 9 o'clock position the whole limbal area glows. The maximum glow will be at the 3 o'clock position. The light from the slit beam is reflected backwards and forwards between the two internal limiting surfaces of the cornea and it is scattered centrifugally all around the cornea.

Filters

The illumination system of many slit lamps incorporates various filters. The blue cobalt filter is used during applanation tonometry (see below) and so is commonly employed.

Both the blue and the green (red-free) filters are useful in examination of the vitreous. The visibility of the gel structure of the vitreous depends upon incident light being scattered. The scattering of light is greatest when the incident light is of short wavelength. Hence blue and green light is scattered more than red light, which is of a longer wavelength.

Another reason for putting up either a blue or green filter when inspecting the vitreous is because under these circumstances there is a relatively dark fundus background. This dark background makes it much easier to detect structures within the vitreous such as the vitreous cortex in posterior vitreous detachment. The reduced background illumination is due to only relatively few rays of shorter wavelengths being reflected from the choroid.

Fundus examination

With the basic slit lamp it is not possible to see further back into the eye than the anterior third of the vitreous. This is because the refractive power of the cornea and lens renders light emerging from the deeper points of the eye parallel. Therefore no image is formed within the focal range of the slit lamp microscope (Fig. 13.35a).

However, the fundus can be rendered visible by using an additional lens to overcome the refractive power of the eye.

The view of the fundus obtained with such a lens is much improved if light reflected from the cornea does not enter the viewing system of the slit lamp microscope. This can largely be achieved if there is no overlap at the cornea of the illuminating and viewing systems.

The illumination column of most slit lamps can be tilted (Fig. 13.33) so that the axis of the illuminating system can be thrown below that of the viewing system (Fig. 13.34).

Fig. 13.33 Slit lamp. To show tilt of illumination system.

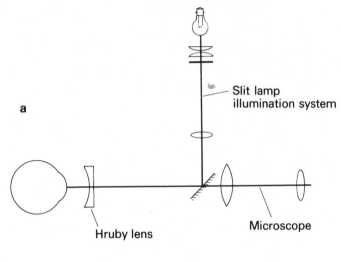

a

Slit lamp illumination system

Hruby lens

Microscope

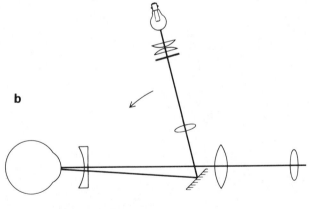

b

Fig. 13.34 Slit lamp. Position of illuminating and viewing axes within the patient's pupil when the illuminating system is tilted.

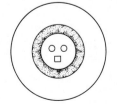

Hruby lens

The Hruby lens is used to examine the fundus and posterior vitreous with the slit lamp microscope. The lens itself is a powerful plano-concave lens, −58.6 D and it is held immediately in front of the eye under examination.

The Hruby lens works by forming a virtual, erect and diminished image of the illuminated retina, the image being anterior to the retina and within the focal range of the slit lamp microscope (Fig. 13.35).

It requires skill to use the lens. The lens is placed with its concave surface towards the eye. The best view is obtained with the lens held near the eye, when the retinal image is found in the pupillary plane.

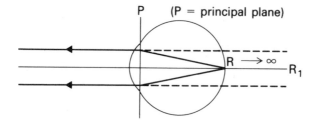

Fig. 13.35 Hruby lens.
(a) Light from the illuminated retina, R, emerges from the eye parallel. The image R_1 of the illuminated retina is formed at infinity and is not accessible to slit lamp examination.

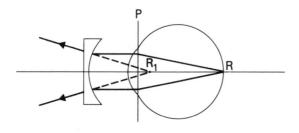

(b) The concave Hruby lens forms a virtual, erect image, R_1, of the illuminated retina, R. The image lies within the focal range of the slit lamp.

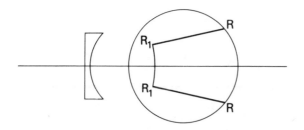

(c) The virtual, erect image R_1 formed by the Hruby lens is also diminished in size, allowing a wide area of retina to be examined without undue rotation of the eye.

Fundus viewing contact lens

The fundus viewing contact lens is a device used to allow examination of the posterior vitreous and posterior pole of the fundus with the slit lamp microscope. The lens in common use is that designed by Goldmann.

The lens is a plano-concave contact lens made of material with a higher refractive index than the eye. When applied to the cornea it allows examination of the fundus by the same mechanism as the Hruby lens.

The central zone of both the gonioscopy lens and the 3-mirror contact lens (see p. 35) may also be used as a fundus viewing lens.

A plano-concave contact lens is used during vitrectomy to allow visualization of the fundus through the operating microscope.

90 D and 78 D lenses

The principle of the indirect ophthalmoscope has been adapted so that the real image of the retina formed by the condensing lens (Fig. 13.10) may be viewed through a slit lamp microscope (Fig. 13.36). For this purpose high power condensing lenses are used, 90 D or 78 D, in order to shorten the light path and bring the retinal image within the focal range of the slit lamp. The loss of image size through using such high power condensing lenses is more than compensated for by the magnification of the slit lamp microscope. The 90 D lens gives a wider field of view but less magnification than the 78 D lens. The technique gives an excellent view of the posterior pole of the fundus but is less suitable for examining the peripheral retina.

Fig. 13.36 90 D (or 78 D) lens.

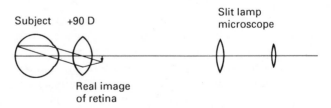

Subject   +90 D

Real image of retina

Slit lamp microscope

Panfunduscope contact lens

A wider field of view is given by the panfunduscope

contact lens (Fig. 13.37). This consists of a high convex power contact lens, which acts as a condensing lens and forms a real, inverted image of the fundus located within a spherical glass element incorporated within the panfunduscope. The spherical glass element flattens the image and redirects the light diverging from it towards the observer. Because the condensing lens is so close to the eye, the field of view is very wide. Indeed the whole fundus as far forward as the equator may be seen in one view without moving the lens. (However, the image formed is correspondingly smaller, and therefore the slit lamp must be adjusted to give high magnification if detailed examination is required.)

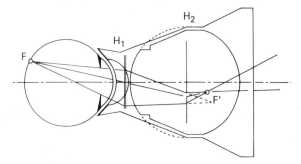

Fig. 13.37 Rodenstock panfunduscope.

Other manufacturers are producing panfunduscope contact lenses which are based on the same principle as the Rodenstock lens, namely that the condensing lens is applied to the eye as a contact lens and the transmitted light collected and redirected towards the observer. The detail of the lens elements varies between manufacturers.

APPLANATION TONOMETER

The applanation tonometer is used to measure the intraocular pressure. The tonometer head is applied to the cornea with sufficient force to produce a standard area of contact. The force required is directly proportional to the intra-ocular pressure when the area of contact is approximately 3 mm in diameter. (When the area of contact is about 3 mm in diameter it can be shown that the effect of surface tension [S] and the rigidity [R] of the cornea cancel each other out [Fig. 13.38]. With larger areas of contact the effect of corneal rigidity becomes appreciable and leads to inaccuracy while with smaller areas of contact the surface tension causes errors.)

Fig. 13.38 Principle of
applanation tonometry.

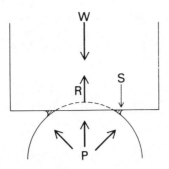

Fig. 13.38 Principle of
applanation tonometry.

When the area of contact is 3.06 mm in diameter, R = S
and W is proportional to P

where W is the force of application of the tonometer.

P is the intra-ocular pressure.

S is the surface tension of the tear/fluorescein ring.

R is the rigidity of the cornea.

Furthermore, when the area of contact is 3.06 mm in
diameter the ocular volume change caused by the
tonometer is very small and does not significantly alter
the intra-ocular pressure. A true reading of intra-ocular
pressure can thus be made.

The Goldmann and Schmidt applanation tonometer
fulfils these conditions, the area of contact being 3.06 mm
in diameter. The instrument consists of the applanation
head which is mounted on a spring-loaded lever. The
tension on the spring is adjustable and the adjustment
knob is calibrated in terms of intra-ocular pressure, in mm
of mercury.

In order to achieve a standard area of corneal contact
the applanation head contains two prisms, mounted with
their bases in opposite directions, Fig. 13.39.

Fig. 13.39 Goldmann and
Schmidt's applanation
tonometer. (Insert shows
alignment of prisms.)

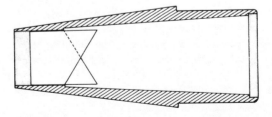

The operator thus looks through the applanation head and
sees the circle of corneal contact split into two half circles
which are laterally displaced in opposite directions by the
prisms (Fig. 13.40). He adjusts the applanation pressure

until the half circles just overlap one another when the area of contact will be exactly 3.06 mm in diameter. (The inner edges of the fluid meniscus define the area of contact and are aligned as shown.)

a Applanation area too small   **b** Applanation area too large   **c** Applanation area correct

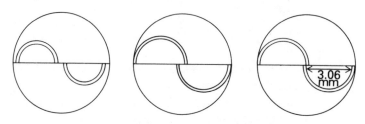

PACHOMETER

The thickness of the cornea and the depth of the anterior chamber may be measured by means of a pachometer. The pachometer uses the Purkinje–Sanson images formed by the anterior and posterior corneal surfaces (images I and II) to measure corneal thickness and the Purkinje–Sanson images formed by the posterior corneal surface and anterior lens surface (images II and III) to measure anterior chamber depth.

The pachometers in current use depend on the principle of image doubling. The doubled image is aligned by the operator so that the surfaces in question (anterior and posterior corneal surfaces, or posterior corneal and anterior lens surface) coincide. The corneal thickness or anterior chamber depth can then be directly read off a scale.

The image doubling can be achieved either by splitting the incident beam of light, or by means of a specially adapted eye-piece which splits the observer's view of the eye. Both types are used in conjunction with a slit lamp.

An example of the first type is the pachometer designed by Maurice and Giardine (1951) (Fig. 13.41).

Fig. 13.40 Goldmann and Schmidt's applanation tonometer. Image seen by operator.

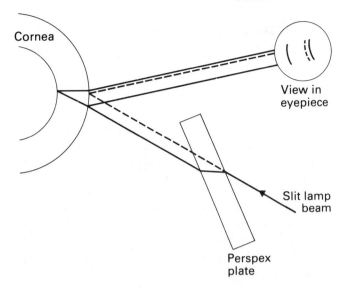

Fig. 13.41 Pachometer of Maurice and Giardine. Measurement of corneal thickness.

A perspex plate covered by coloured celluloid and having a cut-out area is placed in the slit lamp beam. The beam is thus split, some light proceeding undeviated via the cut out zone (dotted line, Fig. 13.41) and some being laterally deviated by passage through the perspex plate. The images formed by the two beams at the surfaces of the cornea are viewed through the slit lamp and the plate rotated until the images in question are superimposed.

The pachometer in current use with the Haag Streit 900 slit lamp depends on principles described by Jaeger. Image doubling is achieved by a specially adapted eye piece which is substituted for the normal slit lamp eye piece. Measurement is made by rotating a transparent plate so that the two images are aligned in such a way that the surfaces in question, e.g. anterior and posterior corneal surfaces are juxtaposed (Fig. 13.42).

Fig. 13.42 Haag Streit (Jaeger) pachometer. Operator's view of eye for measurement of corneal thickness.

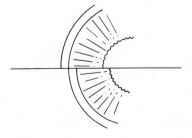

Practical Clinical Refraction

Chapter 14

Having assimilated the optical theory presented in the previous chapters, the practitioner should be able to approach the refraction of a patient understanding the principles of what he is doing. But is this enough? Experienced refractionists know that prescribing is an art as well as a science. There are certain clinical tips that have been handed down over the years, respect for which makes the difference between the mediocre and good refractionist. Some of these will be included in the following sections on practical refraction and are printed in italics.

History

The patient's age and occupation and special visual requirements, e.g. occupation and hobbies, must be ascertained. Visual symptoms, eye-related diseases and family history of eye disease, especially glaucoma, must be enquired for. The patient's past history of spectacle wear is important, and it is wise to find out the previous prescription, if this is available, and to examine the old glasses. *Note the lens form of the previous glasses. Myopes are especially intolerant to a change in lens form and some prefer to continue wearing plano-concave lenses, known as 'flats', even though this may not be the best-form lens for their prescription. Notice the type of bifocal or varifocal lenses in use and whether the patient is happy with them.*

Examination

The visual acuity is measured uniocularly for distance and near, unaided and with existing spectacles. *Fog the fellow eye of patients with nystagmus with a high plus lens as complete occlusion makes the nystagmus worse and lowers the uniocular acuity.* If the acuity is very poor, examine briefly with the ophthalmoscope at this stage to exclude a pathological cause. *More detailed scrutiny of the fundi should be left until after the refraction to avoid photostress-induced reduction of acuity.*

Objective refraction

Perform a cover test to detect any manifest squint. Watch the fellow eye for any movement when one eye is covered. Repeat the manoeuvre after an interval without cover of either eye (to allow refixation of the dominant eye), so that each is covered in turn. If, for example, the uncovered right eye moves outwards to take up fixation when the left eye is covered, a manifest right convergent squint is present. *If the patient has a manifest squint without diplopia, binocular vision is lacking and it will not be possible to test the muscle balance by the Maddox rod and wing tests (see below) which depend on binocular vision.*

Fit the trial frame, taking care that the lens apertures are centred on the pupils with the patient gazing straight ahead. The frame must be level across the face, and as close to the eyes as the lashes will allow. Remember always to place high power trial lenses in the back cell of the trial frame (cf effective power of lenses and back vertex distance, pages 102–105).

Perform the retinoscopy with the patient gazing at a distant object, such as the top letter of the test type. It is important to perform the retinoscopy as close to the patient's visual axis as possible in order to measure the true optical length of the eye. Care should be taken to ensure that the patient's view of the distant fixation object is not obstructed. To this end the examiner should ideally use his right eye to examine the patient's right eye and his left eye for the patient's left eye. If the examiner does get in the way, the patient loses distant fixation and may start to accommodate. *(If the patient does accommodate and fix on the examiner, this will be betrayed because the pupil will constrict and the retinoscopy reflex become more difficult to see.)* During retinoscopy it is considered better to fog the patient's fellow eye rather than to occlude it in order to discourage involuntary accommodation. In the presence of manifest squint the dominant eye may have to be occluded to achieve steady fixation with the non-dominant eye.

In the case of young children or children who have a latent or manifest squint, the ciliary muscle should be paralysed with a topical cycloplegic drug before retinoscopy is performed in order to ascertain the total hyper-metropia present.

Some refractionists place a plus lens of the dioptric value of their working distance (e.g. +1.50 DS for ⅔ metre) in the trial frame before commencing the retinoscopy. Once the point of reversal or neutral point (see page 153) has been reached for both eyes, this lens is removed and the subjective examination started with the lenses remaining in the frame.

After viewing the retinoscopy reflex through the working distance lens, further spherical lenses are added, convex if the movement is 'with' and concave if 'against' (Fig. 14.1a–c) until the neutral point is reached in one meridian (Fig. 14.1d). (If this lies between increments of lens power, reversal of the reflex movement is taken as the end point.) Cylindrical lenses may then be used to neutralise the other meridian and to find the axis of the astigmatism present.

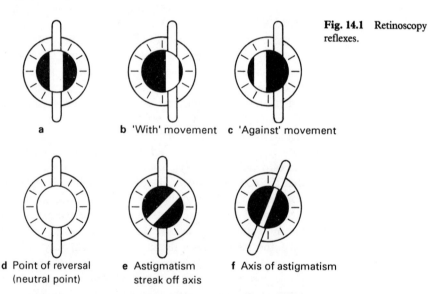

Fig. 14.1 Retinoscopy reflexes.

a

b 'With' movement   c 'Against' movement

d Point of reversal   e Astigmatism   f Axis of astigmatism
   (neutral point)   streak off axis

The axis may be found because the retinoscopy reflex will only align with the axis of the cylindrical lens (and the retinoscopy streak if a streak retinoscope is being used) when they all lie in the axis of the astigmatism. If the axis of the cylindrical lens lies outside the axis of astigmatism, the reflex will move obliquely (Fig. 14.1e). As the point of reversal is approached, it can be shown that the angle of misalignment of the reflex will be six times the angle of

misalignment of the cylindrical lens. The cylinder, therefore, must only be rotated a small amount before recheck-
ing the retinoscopy and this method provides a very
accurate means of determining the axes of astigmatism
objectively.

The same method may also be used to identify the axes
of astigmatism if spherical lenses alone are used during
retinoscopy, e.g. during cycloplegic retinoscopy of a small
child when it is easier to hold lenses in the hand rather than
use a trial frame. The reflex in the pupil will only align
with the streak when both lie in one principal meridian of
astigmatism (Fig. 14.1d–e).

It is often said that minus cylinders should be used
during retinoscopy because the use of plus cylinders when
refracting young hypermetropic patients may stimulate
accommodation as the eye is only fogged in both meridians
when the full plus cylinder is in place. However, if minus
cylinders are always used there is a tendency to overcorrect
hypermetropia in the elderly and when performing cycloplegic refraction. This is because most patients will not
accept the full value of the cylinder found on retinoscopy.
Supposing that at the end of retinoscopy, after correction
for working distance, the trial frame contains a +4.00 DS
and a –1.50 DC. On subjective refraction the patient
tolerates only a –1.00 DC, so the frame now contains
+4.00 DS and –1.00 DC. With this combination of lenses
the lower value meridian is overcorrected by +0.50 D and
the examiner must verify the spherical correction again,
reducing it to +3.50 DS. If, however, the lenses used for
the retinoscopy had been +2.50 DS and +1.50 DC, after
reduction of the cylinder to +1.00 DC, the higher value
meridian would be undercorrected, which is better tolerated, and the spherical correction would not need to be
changed.

Some refractionists therefore prefer to use plus cylinders
with plus spheres, and minus cylinders with minus
spheres. Thus retinoscopy is performed with spheres up to
the lowest meridian of reversal, and continued with cylinders of the same sign to the higher meridian of reversal. The
exception to this is when the initial (uncorrected) retinoscopy reflexes move in opposite directions in the two
meridians, i.e. one 'with' and one 'against'. In this case,
spherical lenses are added until one meridian is neutralised, and cylindrical lenses of opposite sign and higher
power are added until the other meridian is neutralised. *It*

is wise to use minus spheres and plus cylinders for this purpose because it is easier to find the axis of astigmatism using the 'with' movement seen when plus cylinders are being added than the 'against' movement that goes with the use of plus spheres and minus cylinders.

The retinoscopy findings are usually recorded in the UK according to the following convention (Fig. 14.2).

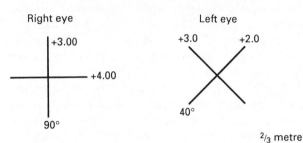

Right eye   Left eye   **Fig. 14.2**   Record of
                                     retinoscopy findings.

A cross is drawn in the orientation of the principal meridians, and the angle of one meridian marked. The dioptric value of the point of reversal is marked on each meridian and the working distance recorded. If this result is transposed into a lens prescription (corrected for working distance) the axis of any cylinder lies at 90° to the line of its meridian, e.g. the left eye in Fig. 14.2 would be

$$\frac{+0.50\,\text{DS}}{+1.00\,\text{DC}}\ \text{axis}\ 40° \qquad \text{or} \qquad \frac{+1.50\,\text{DS}}{-1.00\,\text{DC}}\ \text{axis}\ 130°$$

In practice retinoscopy is not always easy.

In cases of high refractive error, the initial reflex may be too dim and diffuse to be identified. Retinoscopy should be repeated using a moderately strong convex and concave lens, e.g. + or –7.00 DS, one of which should bring the reflex into view, assuming that the media are clear. Alternatively, examination of the eye with the direct ophthalmoscope will allow differentiation between high myopia and high hypermetropia.

If the refraction varies between the central and peripheral parts of the pupillary aperture there may be an increase in brightness in the centre or periphery due to spherical aberrations. For example, in nucleosclerosis the central zone is relatively myopic compared to the periphery and the centre of the pupil appears bright. To judge the end-point, which may not be as sharply defined as usual, concentrate on the central zone in which the visual axis lies.

It is sometimes difficult to decide whether the reflex is moving 'with' or 'against' the movement of the retinoscope because there appear to be two reflexes in the pupil, moving towards and away from each other like the blades of a pair of scissors, so-called 'scissor shadows' (Fig. 14.3a). This is due to a difference in refraction between different zones of the pupillary aperture. It is more commonly seen with a dilated pupil and near the end-point of retinoscopy, when one area will be relatively myopic, M, while the other is relatively hypermetropic, H, to the plane of observation, R (Fig. 14.3b). It is traditionally taught that the end-point is taken when the two reflexes meet in the centre of the pupil, but this may be difficult to judge. In our experience, one blade of the scissors is brighter than the other and the end-point is taken when the brighter reflex reverses.

Fig. 14.3a Retinoscopy — scissor shadows.

Fig. 14.3b Retinoscopy — optical diagram to show cause of scissor shadows.

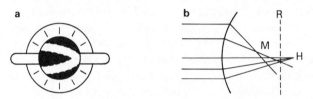

Retinoscopy may be very difficult in keratoconus, when a characteristic swirling reflex is seen and the reflex from the apex of the cone is darker than the periphery (the oil-drop sign). It may be impossible to detect an end-point and subjective testing must then be relied upon.

Subjective refraction

Make the correction for your working distance (add –1.50 DS for ⅔ metre or remove the 'working distance lens' if one was used) *and, because patients usually do not tolerate the full cylindrical correction, it helps to reduce the cylinder by approximately a quarter of its value (see above)* before commencing the subjective examination.

Occlude the fellow eye (unless nystagmus is present, in which case use a fogging technique, page 172). Using the distance test type, first verify the sphere by offering small plus and minus additions until no further improvement can be made. Patients with good visual acuity can appreciate a difference of 0.25 DS while those with poor acuity may only appreciate larger increments, e.g. 0.50 DS. Next verify the axis of the cylinder before adjusting its power (cf the cross cylinder, Chapter 6, page 63). If a large change is found in the cylinder, it is wise to go back and recheck the sphere. If the patient is myopic, the duochrome test (Chapter 8, page 77 should be done monocularly and binocularly.

If the red letters are only marginally clearer than the green when the patient views them monocularly, the green letters may be clearer when the eyes are used together. A small reduction, e.g. +0.25 DS to one or both eyes may be needed to make the red letters clearer binocularly and ensure that the patient will be comfortable, and not accommodating, when wearing the prescription. Record the prescription and acuity for each eye, and the binocular acuity.

Use the Maddox rod (Chapters 6, page 59) to check the muscle balance for distance. *Some patients are initially unable to see the white spot light and the red line, especially if there is confusing side illumination in the room. Occlude each eye in turn to ascertain that the spot and line are visible to the appropriate eye uniocularly, and then uncover both eyes and see if the patient is able to perceive them simultaneously. If the patient sees them one at a time uniocularly but not simultaneously with both eyes uncovered, his binocular vision must be defective or absent.* In cases where the Maddox rod test reveals a significant extraocular muscle imbalance, especially a vertical one, check to see if the addition of the appropriate prism improves the patient's binocular comfort and acuity. *The full value of prism as found by the Maddox rod test is rarely required.*

If the patient is presbyopic, calculate the likely reading addition and add this to the distance lenses in the trial frame. (In practice the reading addition is estimated by rule-of-thumb from the patient's age:

45–50 years = +1.00 D addition
50–55   = +1.50 D addition
55–60   = +2.00 D addition
over 60   = +2.00 D addition)

Beware of prescribing too great a reading addition (cf Chapter 11). *(The most frequent reason that patients seek a retest is that too strong a near addition has been prescribed. In normal circumstances not more than +2.50 DS addition should be given.)* Record the near acuity for each eye alone and binocularly.

Use the Maddox wing to check the near muscle balance.

Orthophoria for distance but a large exophoria for near indicates convergence insufficiency. This may cause such symptoms as headache or eye-strain after close work, or it may be asymptomatic. If the patient has symptoms, the convergence should be strengthened by means of convergence exercises. *If base-in prisms are prescribed the convergence may become weaker still, and progressively stronger prisms will be required.* Vertical muscle imbalances for near may require prismatic correction, but again the full value of prism is rarely required or tolerated.

Measurement of interpupillary distance

It is sometimes useful to measure the patient's interpupillary distance, although this is usually done in the UK by the dispensing optician. The measurement of the interpupillary distance is important in babies and small children, especially if high power lenses have been prescribed, as decentration caused by an unsuitable spectacle frame will introduce an unwanted prismatic effect (cf spherical lens decentration and prism power, pages 56–7). Poor centration of aphakic spectacle lenses also causes an unwanted prismatic effect and this is a common cause of intolerance of aphakic glasses.

The anatomical interpupillary distance may be measured by the following methods. A millimetre rule is rested across the bridge of the patient's nose and the patient asked to look at the examiner's right eye. The zero of the rule is aligned with the nasal limbus of the patient's left eye (which will be looking straight ahead at the examiner's right eye). The patient is then instructed to look at the examiner's left eye and the position of the temporal limbus of the patient's right eye is noted, giving the anatomical interpupillary distance. (The measurements are taken from limbus to limbus to exclude inaccuracy due to differences or changes in pupil size.) Alternatively, a fixation light may be held in front of each of the examiner's eyes in turn and a similar procedure followed, the distance between the corneal light reflexes on the patient's eyes being measured.

The measurement of importance in making spectacles is the distance between the visual axes for distance vision which is approximately 1 mm less than the anatomical interpupillary distance.

Helpful hints to avoid intolerance

Do not change the axis of the cylinder, especially in a myope, unless there are compelling reasons for doing so.

Do not change the lens form worn by a myope.

Do not prescribe a large cylinder for any patient who has never worn a cylindrical correction before. Break them in gradually. The exception to this is pseudophakic patients who tolerate a full astigmatic correction well.

Do not overcorrect hypermetropes, better to leave them 0.25 DS undercorrected so they can read the bus numbers in the far distance.

Do not fully correct myopes, better to leave them 0.25 DS undercorrected so they do not have to use their accommodation for distance.

Do not give too great a reading addition, so the patient cannot read the newspaper held at arms' length (cf pages 121, 178).

Do not recommend bifocal or progressive lenses without carefully considering the needs, occupation and frailties of the patient.

Discuss with the patient the subjective and practical points relevant to the new prescription, e.g. warn a new bifocal wearer to be careful at steps.

Do not alter a satisfactory prescription unless there is a very definite reason to do so.

Do not advise the patient to buy new glasses because of a minor change in prescription that he will not be able to appreciate subjectively.

Notes on the management of refractive errors in children

The refraction and management of children with refractive errors is possibly the most interesting and rewarding aspect of the refractionist's work. The following discussion relates to the management of the age-group at risk of developing squint or amblyopia, e.g. birth to approximately 8 years of age.

As stated above, cycloplegic refraction is necessary to obtain an accurate retinoscopy in infants and children. If this is done, some degree of ametropia will be found in most cases and judgement must be exercised to decide whether spectacles will be beneficial. (Most parents hope their children will not need glasses, and the children tolerate glasses better if they perceive that they see better when wearing them.)

Hypermetropia is very common in infants and children and requires a full correction in the presence of esophoria or esotropia. It is also wise to correct it if there is a family history of esotropia. Moderate or high degrees of hypermetropia require correction to achieve good visual acuity, and more than 1 dioptre of hypermetropic anisometropia must be corrected to prevent refractive amblyopia in the more hypermetropic eye, (cf page 97).

The uncorrected hypermetropic child overcomes some or all of his hypermetropia by exercising extra accommodation. When glasses are worn for the first time, the accommodation may not relax and the vision will be blurred. Usually the accommodation relaxes after a few days if the glasses are persevered with, but it is wise to warn the parents about this when the glasses are prescribed. Occasionally the accommodation fails to relax and the prescription needs to be reduced for a while, or a short course of cycloplegic drops may do the trick.

Myopia of sufficient degree to prevent the child from seeing what is written on the blackboard should be corrected before the child starts formal schooling. Myopia should also be corrected in the presence of exophoria or exotropia in order to stimulate accommodation and convergence.

It is wise to explain to the parents that the glasses will have to be made stronger as the child grows bigger, because the eye also grows, and this progression of the myopia should not be a cause of undue concern, and will stabilise when growth stops.

High myopia should be corrected as soon as it is detected because it impairs the child's perception of the environment and may retard development in many fields, e.g. mobility, relating to other people, recognising shapes and objects and learning to use the hands, etc. *Experience shows that in both adults and children a full correction is not tolerated in high myopia. Such patients should be left 1 or 2 dioptres undercorrected.*

Uniocular moderate or high myopia often causes amblyopia and an attempt should be made to correct it. Surprisingly good results can be achieved with spectacles, if started early, even in high degrees of anisometropia as the young brain seems to cope with much greater degrees of aniseikonia than the adult is able to. Sometimes a contact lens for a unilateral highly myopic eye will be effective where glasses have failed, but the age and suitability of the child and family must be taken into account as contact lens wear entails practical problems, especially in children.

Astigmatism needs correction if the visual acuity is reduced in the affected eye. Most babies and many infants have quite high degrees of astigmatism but in many cases this reduces or disappears by the age of $2\frac{1}{2}$ or 3 years when the first accurate tests of visual acuity become possible. Astigmatism alone rarely needs correction in infants. In children of 3 years of age and older, astigmatism of 1 dioptre or more, which is sufficient to impair the visual acuity, should be corrected.

Finally, remember to check the fit of the frames for sore spots or chafing if a child refuses to keep his glasses on. It is worth telling the parents to do this when new glasses are acquired as non-spectacle wearers may not think of this as a cause of non-tolerance.

Chapter 15 **Automated Clinical Refraction**

Over the last 200 years or so attempts have been made to automate the process of refraction, but with little success. No reliable substitute could be found for the skilled human refractionist. Recently, a new generation of autorefractors has appeared on the market and it is therefore important to understand the underlying principles on which they function as well as the difficulties which prevented the successful automation of refraction in the past.

Basic principles used in automated refraction

The Scheiner principle

Scheiner discovered in 1619 that the point at which an eye was focused could be precisely determined by placing double pinhole apertures before the pupil of the eye. Parallel rays of light from a distant object are reduced to two small bundles of light by the Scheiner disc (Fig. 15.1). These form a single focus on the retina if the eye is emmetropic (Fig. 15.1a) but if there is any refractive error two spots of light fall on the retina (Figs 15.1b–c). By adjusting the position of the object (mechanically or optically) until one focus of light is seen by the patient, the far point of the patient's eye and the refractive error can be determined. It is generally considered that the judgement of singleness of image gives a more precise end-point than the judgement of least blur of an image.

By this method the eye is examined only along the paths of the two bundles of light transmitted by the Scheiner disc. This is the earliest of a class of 'zonal focus' methods of refraction in which the overall refractive condition is determined by examining through small zones of the optical aperture. The zonal focus methods based on the Scheiner principle have been widely used in the design of automated refractors.

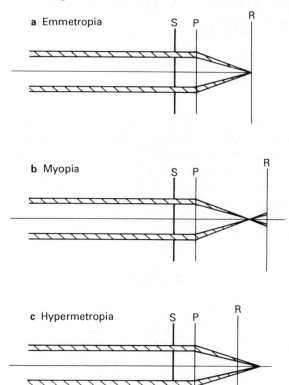

a Emmetropia

b Myopia

c Hypermetropia

Fig. 15.1   Scheiner principle.

The optometer principle

The term 'optometer' was first used in 1759 by Porterfield who described an instrument for 'measuring the limits of distinct vision, and determining with great exactness the strength and weakness of sight'. The principle of the optometer is illustrated in Fig. 15.2.

A convex lens is placed in front of the eye so that its focus lies in the spectacle plane and a movable target is viewed through the lens (Fig. 15.2a). If the target lies at the first principal focus of the lens, light from the target will be parallel at the spectacle plane, and focused on the retina of the emmetropic eye (Fig. 15.2b). When the target is within the focal length of the lens, light from it will be divergent in the spectacle plane (simulating a concave trial lens) (Fig. 15.2c) while light from a target outside the focal length of the lens will be convergent in the spectacle plane (simulating a convex trial lens) (Fig. 15.2d). The vergence

Fig. 15.2 Optometer principle
and optometer.

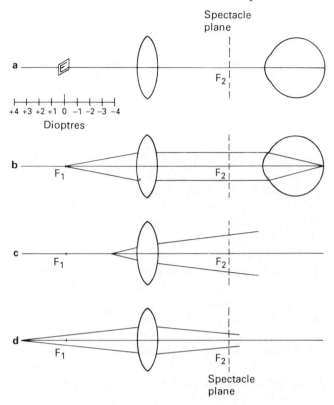

of the light in the plane of the second principal focus of the
lens is linearly related to the distance of the target from the
first principal focus of the lens. The instrument is calibrated to show the vergence of the light in the focal plane,
in dioptres, according to the position of the target (Fig.
15.2a).

Meridional refractometry

In the presence of astigmatism, the axes of the principal
meridians must be found and the refraction in both
measured. This can be done even with the early instruments as long as the examiner plays an active part in the
process. However, the need to identify the principal
meridians of astigmatism stood in the way of truly automated refraction until the principle of meridional refractometry was discovered in the 1960s.

It was realised that if the spherical refraction is measured in at least three arbitrary meridians, the position of

the principal axes and their refractive power can be found by mathematical calculation. Greater accuracy can be achieved if measurements are taken in more than three arbitrary meridians, although the mathematics becomes more complicated.

Early optometers

The earliest instruments were the subjective optometers. The patient had to adjust the instrument to achieve the best subjective focus or alignment of a target. However, the subjective optometers proved unsatisfactory because of alignment problems, irregular astigmatism and instrument accommodation.

Alignment is critical in a system based on the Scheiner principle, as both pinhole apertures must fall within the patient's pupil. Achieving and maintaining correct alignment of the instrument requires great skill and patience from the examiner and good cooperation from the patient.

Instruments using the Scheiner principle measure only the refraction of two small portions of the pupillary aperture corresponding to the apertures in the Scheiner disc. Irregular astigmatism is present to some degree in all eyes, but if it is present to a significant extent the overall refraction of the eye may be very different from the result obtained by this method.

Inappropriate accommodation often occurs when a target is viewed which is known to be within an instrument, and therefore near the eye. This is called instrument accommodation, and it has been a major problem in optometer design. Many ingenious devices have been employed in an attempt to overcome the tendency of the patient to involuntarily accommodate, but it remains a problem even in some of the modern instruments. The degree of instrument accommodation often fluctuates during the measurement process, introducing error in the astigmatic as well as the spherical correction found if the meridians are not measured simultaneously.

Between the two world wars several so-called objective optometers were developed, but these required the examiner to focus or align the image of a target on the patient's retina, so were not truly objective. They failed to come into general use because of alignment difficulties and instrument accommodation, but three of these instruments are still available and have gained acceptance in Europe where they are used in place of retinoscopy. (These are the Rodenstock Eye-Refractometer, the Hartinger Coincidence Refractionometer and the Topcon Refractometer.)

Infrared optometers

In recent years, the automatic infrared optometers have come to the fore. These are truly objective instruments as the instrument itself senses the end-point of refraction. Their development has sprung from the advances in electronics and microcomputers that have taken place in recent decades, and has been facilitated by the discovery of the principle of meridional refractometry.

The instruments commercially available filter out all but infrared light from the measuring system and detect the end-point by means of an electronic focus detector. Some are based on the Scheiner principle, some simulate retinoscopy and others use the optometer principle.

The patient's eye is refracted using invisible infrared light to overcome instrument accommodation. (A separate fixation target must still be provided, and is designed to encourage relaxation of accommodation.) However, because of the chromatic aberration of the eye and because infrared light is not reflected by the same layers of the retina as visible light, the refraction of the eye to infrared differs significantly from its refraction to visible light. This difference is of the order of 0.75 D to 1.50 D more hypermetropic to infrared, and may vary slightly from one individual to another. Manufacturers therefore calibrate the instruments empirically to correlate with subjective clinical results.

Furthermore, the instruments do not perform well if the eye has a small pupil or distorted pupil, e.g. broad iridectomy, or if the ocular media are not clear (as a guideline, they become inaccurate if media haze exceeds that which would reduce the vision to 6/18).

Other recent developments

With the proliferation of modern electronics and technology many new means of assessing the refraction of the eye have been developed.

These include *photographic* methods in which the retinoscopic image of one or more point sources of light is photographed and studied later to estimate the refraction. So far these methods have only found limited use in screening for refractive errors and are not sufficiently accurate to be used for the prescription of spectacles.

Laser speckle pattern refraction is a subjective method in which the patient views the interference patterns produced by reflecting a laser beam off a slowly rotating drum. To the patient the drum appears to be covered with red and black speckles which move in one direction or the other depending on whether the patient's far point lies in front of or beyond the surface of the drum. This method is sensitive to 0.25 D in a given meridian, but as yet no method has been devised to measure the axis of any astigmatism.

Sophisticated subjective refracting systems are coming onto the market, but they require a skilled operator and depend on the subjective responses and cooperation of the patient. One such system is the *Humphrey Vision Analyser* in which the patient views targets imaged in a concave mirror 3 metres in front of him. Separate targets are presented to each eye by the machine and the vergence of light from these can be so adjusted that a full binocular subjective refraction can be done including near vision and muscle balance. The focus of the targets is adjusted by means of Alvarez variable-power lenses. The Alvarez lens is based on the fact that any refractive power can be achieved by the sum of a variable sphere, a variable cross cylinder axes 90° and 180°, and a variable cross cylinder axes 45° and 135°. Each Alvarez lens unit consists of two lenses with complicated optical surfaces which fulfil the above requirements and which can be moved with respect to each other along two mutually perpendicular axes in the plane of the lenses. When the lenses are exactly opposite each other their combined power is zero. Any spherocylindrical power may be obtained by moving them the appropriate amount with respect to each other. The Vision Analyser enables subjective refraction to be performed without any apparatus on or close to the patient's face, but it does require a skilled operator and cooperative patient.

At the end of the day there is so far no machine that can equal the experienced refractionist in accuracy, ability to test abnormal eyes, patience with the very young and elderly, and wisdom to prescribe the correction that will be most suitable and acceptable to the patient.

Appendix

Mathematical definitions

1. Sine, Cosine and Tangent.

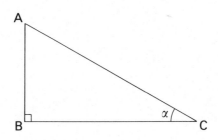

$$\text{Sine } \alpha = \frac{AB}{AC}$$

$$\text{Cosine } \alpha = \frac{BC}{AC}$$

$$\text{Tan } \alpha = \frac{AB}{BC}$$

2. Angles formed by parallel lines

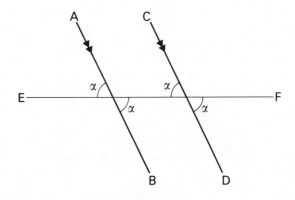

The angles α formed by parallel lines AB and CD intersecting line EF are all equal.

Index